Essential Mathematics for Quantum Computing

A beginner's guide to just the math you need without needless complexities

Leonard S. Woody III

BIRMINGHAM—MUMBAI

Essential Mathematics for Quantum Computing

Publishing Product Manager: Sunith Shetty

Senior Editor: Nathanya Dias

Content Development Editor: Sean Lobo

Technical Editor: Rahul Limbachiya

Copy Editor: Safis Editing

Project Coordinator: Aishwarya Mohan

Proofreader: Safis Editing

Indexer: Manju Arasan

Production Designer: Nilesh Mohite

Marketing Coordinator: Abeer Dawe

First published: April 2022

Production reference: 1170322

Published by Packt Publishing Ltd.

Livery Place

35 Livery Street

Birmingham

B3 2PB, UK.

ISBN 978-1-80107-314-1

www.packt.com

To my wife Jeanette, I owe you a debt of gratitude that I can only repay by loving you every day for the rest of my life, and fortunately for us, that will be easy.

I dedicate this book to my mom, Georgia Chandler Mapes, and my dad, Leonard Spencer Woody, Jr.

You raised me right!

And to my grammy, Patricia Dana Woody. You were my second mother and I love you and miss you terribly.

Acknowledgements

I would first like to acknowledge my technical reviewer, Emmanuel Knafo, Ph.D. He spent tireless hours reviewing this text and it would not be the book it is without him. Secondly, I would like to thank my close friend Sam Smith, who reviewed many chapters quickly and eagerly. Sam, Robin Smith, Rory Woods, and I came up together at Microsoft. Thank you for your friendship, our many happy hours, and help with the book. My first manager at Microsoft, my friend and mentor Omar Kouatly, allowed me to get started in this venture of quantum computing, encouraged me, and helped with the book as well. Thank you. Delbert Murphy, Darius Zakrzewski, and Jon Skerrett have been my "partners in crime" in exploring, learning, and sharing a passion for quantum computing. Thank you for your inspiration. Finally, my friend Matthew A. Kirsch helped with early copies of this text and earlier parts of my life. I thank you for those immeasurable contributions as well.

In the one year plus that it took to write this book, I needed support and advice. My great friend and spiritual mentor, Art Thompson, provided that in spades. Other close friends such as Graham Eddy, Carmel Maddox, Heather Downey, Patrick Sweet, Eli Rosenblatt, Rich Chetelat, Paul Varela, Benjamin Maddox, Nacho Dave, and Andy Brown have been there every step of the way during this tumultuous year.

No book is written alone and I would like to thank the people at Packt for working with me to make this book a reality. I would especially like to thank Sean Lobo, my editor, for sticking with me all the way through and his many hours spent reviewing this text.

Finally, I would like to thank my family, which includes Brandi Zahir and her children Zachary, Benjamin, and Caitlyn. To my children, you allowed me to write this book and gave up many hours with daddy so that I could finish it. I will always love you and you are the reason I exist. Thank you, Eva-Maria, Sophia, Johnny, and Alex. To my wife Jeanette of 17 years, you are the love of my life, my rock, my person. We have built quite a family together and I can't wait to live the rest of my life with you. And to who made this all possible, thank you, God.

Contributors

About the author

Leonard S. Woody III is a senior consultant with 20 years of experience explaining complex subjects to software development clients. For the last 3 years, he has worked at Microsoft, most currently as a program manager for Azure Quantum. He was awarded a BS in computer science and a BS in physics from the University of Virginia. He attained his MS in software engineering from George Mason University. Woody lives in Northern Virginia with his wife and four children. His biggest love is spending time with his family.

About the reviewers

Emmanuel Knafo, focusing on DevOps innovation and cloud architecture, helps organizations transform how to ideate, plan, execute, and learn from their technology investments. He obtained his Ph.D. in mathematics in number theory at the University of Toronto. He is a published author in various mathematical journals. He has published IT articles on the Microsoft Premier Developer Blog.

I would like to thank the author for this opportunity to re-ignite my passion for mathematics and physics by making me the technical reviewer for this book. It has been a thoroughly enjoyable experience! My passion for math was instilled by my father, Emile, and nurtured by my mother, Evelyne. Finally, I'm grateful to Audrey and the lights of our lives: Ethan and Adam.

Devika Mehra started her programming journey when she was 15 years old, which led to her never-ending zest to explore the boundless field of technology. She has an immense interest in the fields of security and quantum computing. She initially flexed her muscles in different programming languages and then focused on the development of Android applications. She is currently working with Microsoft Sentinel as a software engineer and develops security integration and analysis content for the end customer. She wishes to make the world a better place to live in and believes that technology can be a great catalyst to achieve this.

Srinjoy Ganguly works as a quantum AI research scientist at Fractal Analytics. He has 4+ years of experience in quantum computing, and is an IBM Qiskit advocate and educator. He also teaches quantum computing at Woxsen University as a visiting professor. His research interests include QNLP, category theory with compositionality, variational quantum algorithms and their applications, and machine learning.

Table of Contents

Section 2: Elementary Linear Algebra

3
Foundations

4
Vector Spaces

5
Using Matrices to Transform Space

Section 3: Adding Complexity

6
Complex Numbers

7

EigenStuff

8

Our Space in the Universe

Appendix 3
Trigonometry

Appendix 4
Probability

Appendix 5
References

Preface

This book is written for software developers and tech enthusiasts that have not learned the math required for quantum computing either in many years or possibly not at all. Quantum computing is based on a combination of quantum mechanics and computer science. These two subjects, quantum mechanics and computer science, are built on a foundation of math, as the following diagram illustrates:

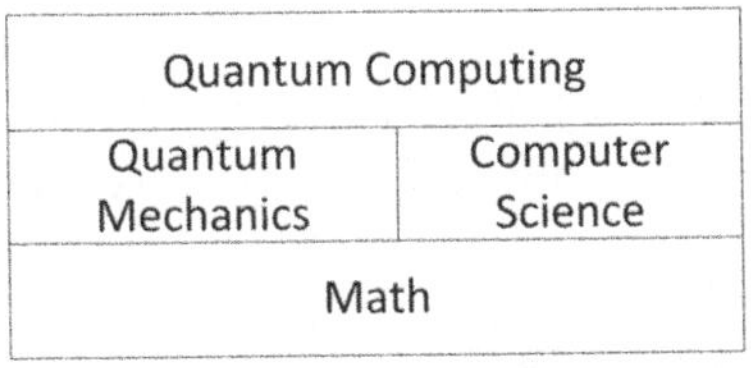

Quantum Computing
Quantum Mechanics | Computer Science
Math

Figure 1 – Diagram of relationship of math to quantum computing

Making sure your foundation is well built as you dive into quantum computing is paramount to your long-term success in the field. Notice that I said "as you dive" instead of "before you dive," because you should do cool quantum computing stuff as you are learning the relevant math. We do that in the very first chapter, *Chapter 1, Superposition with Euclid*, and almost every chapter after that. It's important that you see how the math connects to actual quantum computing.

How to use this book

Let's answer the question of how you should learn the math of quantum computing using this book. Everyone is different, but the steps we used in school work really well (with a few tweaks):

1. **READ THIS AWESOME BOOK!**
2. **Watch YouTube videos** – here are a couple of playlists to get you started:
   - *Essence of linear algebra by 3Blue1Brown* (`https://tinyurl.com/233ruczb`)
   - *Linear Algebra: An In-Depth Introduction by MathTheBeautiful* (`https://tinyurl.com/464dvc4b`)

3. **Exercise** – do actual math problems. You don't learn a sport by just reading a book and watching some videos. You have to do it! This book has you covered with exercises in every chapter.

I'd like to add three more steps to make sure you don't lose enthusiasm:

1. Go do some cool quantum computing stuff with your newfound knowledge.
2. Get stuck.
3. Come back to the book and start again at *Step 1* to get unstuck.

   Now I'd like to quickly talk about who this book is for.

Who this book is for

I don't assume much in terms of mathematical training. A general high school study of math is all you need and even then, I include appendices to review subjects such as trigonometry if you need it. The most important prerequisite is an enthusiasm for quantum computing.

I'll quickly say this book is not for graduate students, mathematicians, physicists, and rocket scientists in general. You probably know all this stuff already. But it might show you a new way to teach it.

What this book is not

This is not an overall introduction to quantum computing. We will most certainly connect the math to quantum computing and do some actual quantum computing. But in the end, this is a math book. There are great books out there that give a general introduction to quantum computing. A personal favorite of mine is *Quantum Computing Explained* by David McMahon and there are more included in the appendix of references I have used for the book.

Download the color images

We also provide a PDF file that has color images of the screenshots and diagrams used in this book. You can download it here: `https://static.packt-cdn.com/downloads/9781801073141_ColorImages.pdf`.

Conventions used

There are a number of text conventions used throughout this book.

Bold: Indicates a new term, an important word, or words that you see onscreen. For instance, words in menus or dialog boxes appear in **bold**. Here is an example: "Select **System info** from the **Administration** panel."

> **Tips or Important Notes**
> Appear like this.

Get in touch

Feedback from our readers is always welcome.

General feedback: If you have questions about any aspect of this book, email us at `customercare@packtpub.com` and mention the book title in the subject of your message.

Errata: Although we have taken every care to ensure the accuracy of our content, mistakes do happen. If you have found a mistake in this book, we would be grateful if you would report this to us. Please visit `www.packtpub.com/support/errata` and fill in the form.

Piracy: If you come across any illegal copies of our works in any form on the internet, we would be grateful if you would provide us with the location address or website name. Please contact us at `copyright@packt.com` with a link to the material.

If you are interested in becoming an author: If there is a topic that you have expertise in and you are interested in either writing or contributing to a book, please visit `authors.packtpub.com`.

Share Your Thoughts

Once you've read *Essential Mathematics for Quantum Computing*, we'd love to hear your thoughts! Scan the QR code below to go straight to the Amazon review page for this book and share your feedback.

`https://packt.link/r/1-801-07314-7`

Your review is important to us and the tech community and will help us make sure we're delivering excellent quality content.

Section 1: Introduction

This section starts the book off with easy concepts such as vectors and matrices.

The following chapters are included in this section:

- *Chapter 1, Superposition with Euclid*
- *Chapter 2, The Matrix*

1
Superposition with Euclid

Mathematics is the language of physics and the foundation of computer science. Since quantum computing evolved from these two disciplines, it is essential to understand the mathematics behind it. The math you need is linear in nature, and that is where we will start. By the time we are done, you will have the mathematical foundation to fundamentally understand quantum computing. Let's get started!

In this chapter, we are going to cover the following main topics:

- Vectors
- Linear combinations
- Superposition

Vectors

A long time ago in a country far, far away, there lived an ancient Greek mathematician named Euclid. He wrote a book that defined space using only three dimensions. We will use his vector space to define superposition in quantum computing. Don't be fooled—vector spaces have evolved tremendously since Euclid's days, and our definition of them will evolve too as the book progresses. But for now, we will stick to real numbers, and we'll actually only need two out of the three dimensions Euclid proposed.

To start, we will define a **Euclidean vector** as being a line segment with a length or magnitude and pointing in a certain direction, as shown in the following screenshot:

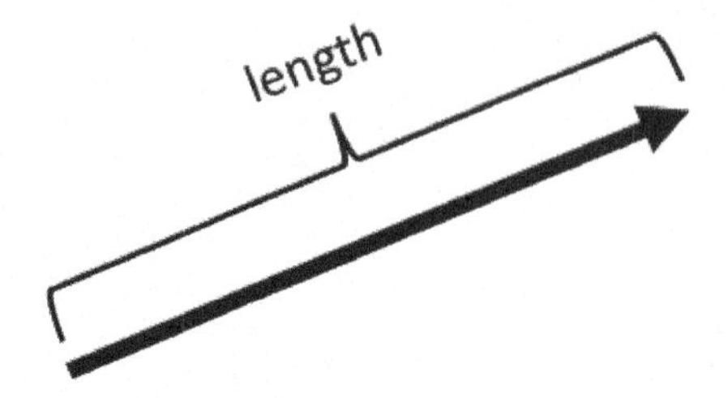

Figure 1.1 – Euclidean vector

Two vectors are equal if they have the same length and direction, so the following vectors are all equal:

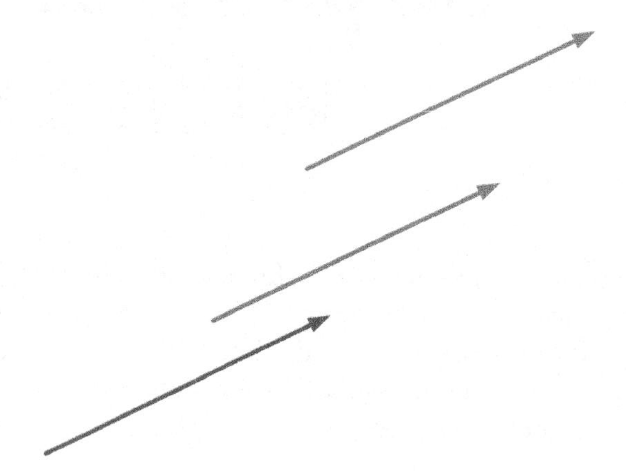

Figure 1.2 – Equal vectors

Vectors can be represented algebraically by their components. The simplest way to do this is to have them start at the origin (the point (0,0)) and use their x and y coordinates, as shown in the following screenshot:

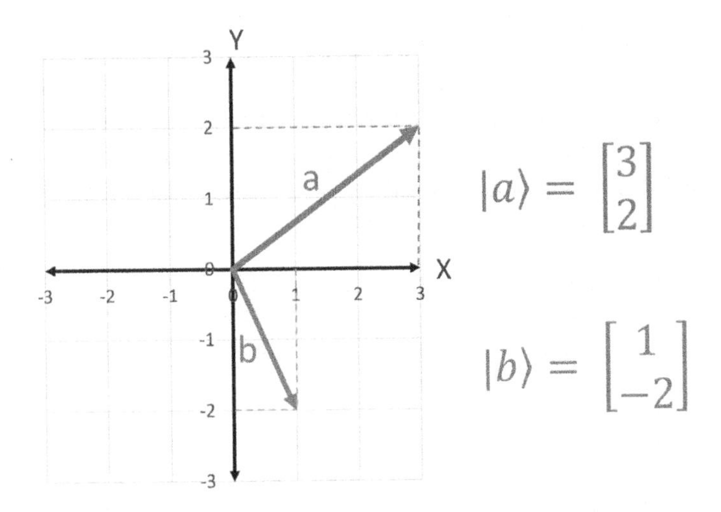

Figure 1.3 – Vectors represented geometrically and algebraically

You should note that I am using a special notation to label the vectors. It is called **bra-ket notation**. The appendix has more information on this notation, but for now, we will use a vertical bar or pipe, |, followed by the variable name for the vector and then an angle bracket, ⟩, to denote a vector (for example, $|a\rangle$). The coordinates of our vectors will be enclosed in brackets [ ]. The *x* coordinate will be on top and the *y* coordinate on the bottom. Vectors are also called "kets" in this notation—for example, ket a, but for now, we will stick with the name vector.

Vector addition

So, it ends up that we can add vectors together both geometrically and algebraically, as shown in the following screenshot:

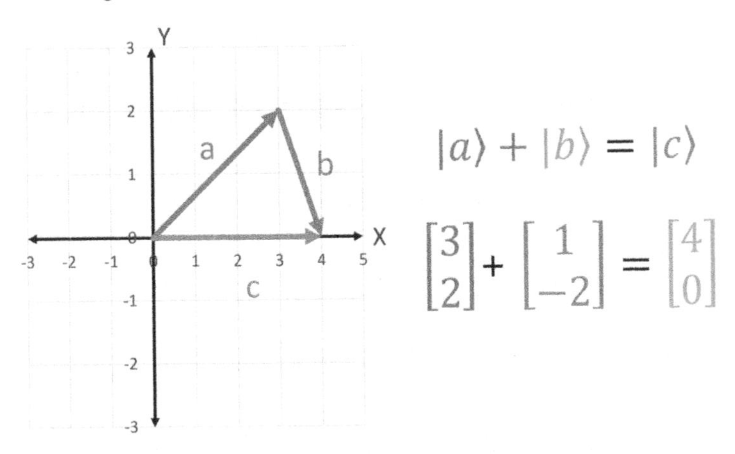

Figure 1.4 – Vector addition

As you can see, we can take vectors and move them in the XY-plane as long as we preserve their length and direction. We have taken the vector $|b\rangle$ from our first graph and moved its start position to the end of vector $|a\rangle$. Once we do that, we can draw a third vector $|c\rangle$ that connects the start of $|a\rangle$ and the end of $|b\rangle$ to form their sum. If we look at the coordinates of $|c\rangle$, it is four units in the x direction and zero units in the y direction. This corresponds to the answer we see on the right of *Figure 1.4*.

We can also do this addition without the help of a graph, as shown on the right of *Figure 1.4*. Just adding the first components (3 and 1) gives 4, and adding the second components of the vectors (2 and -2) gives 0. Thus, vector addition works both geometrically and algebraically in two dimensions. So, let's look at an example.

Example

What is the sum of $|m\rangle$ and $|n\rangle$ here?

$$|m\rangle = \begin{bmatrix} 5 \\ 1 \end{bmatrix} \quad |n\rangle = \begin{bmatrix} 2 \\ -4 \end{bmatrix}$$

The solution is:

$$|m\rangle + |n\rangle = \begin{bmatrix} 5 \\ 1 \end{bmatrix} + \begin{bmatrix} 2 \\ -4 \end{bmatrix} = \begin{bmatrix} 7 \\ -3 \end{bmatrix}$$

Exercise 1

Now, you try. The answers are at the end of this chapter:

- What is $|m\rangle$ - $|n\rangle$?
- What is $|n\rangle$ - $|m\rangle$?
- Solve the following expression (notice we use **three-dimensional** (**3D**) vectors, but everything works the same):

$$\begin{bmatrix} 3 \\ -2 \\ 1 \end{bmatrix} + \begin{bmatrix} 5 \\ 3 \\ -3 \end{bmatrix}$$

Scalar multiplication

We can also multiply our vectors by numbers or scalars. They are called **scalars** because they "scale" a vector, as we will see. The following screenshot shows a vector that is multiplied by a number on the left and the same thing algebraically on the right:

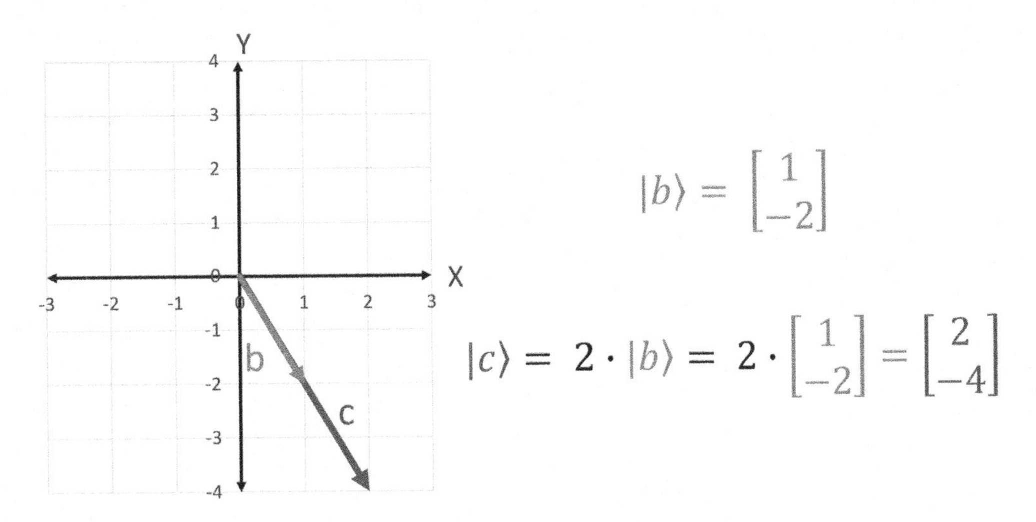

Figure 1.5 – Scalar multiplication

The vector $|b\rangle$ is doubled or multiplied by two. Geometrically, we take the vector $|b\rangle$ and scale its length by two while preserving its direction. Algebraically, we can just multiply the components of the vector by the number or scalar two.

Example

What is triple the vector $|x\rangle$ shown here?

$$|x\rangle = \begin{bmatrix} -4 \\ 2 \end{bmatrix}$$

The solution is:

$$3|x\rangle = 3 \cdot \begin{bmatrix} -4 \\ 2 \end{bmatrix} = \begin{bmatrix} -12 \\ 6 \end{bmatrix}$$

Exercise 2

- What is $4|x\rangle$?
- What is $-2|x\rangle$?

Linear combinations

Once we have established that we can add our vectors and multiply them by scalars, we can start to talk about linear combinations. Linear combinations are just the scaling and addition of vectors to form new vectors. Let's start with our two vectors we have been working with the whole time, $|a\rangle$ and $|b\rangle$. I want to scale my vector $|a\rangle$ by two to get a new vector $|c\rangle$, as shown in the following screenshot:

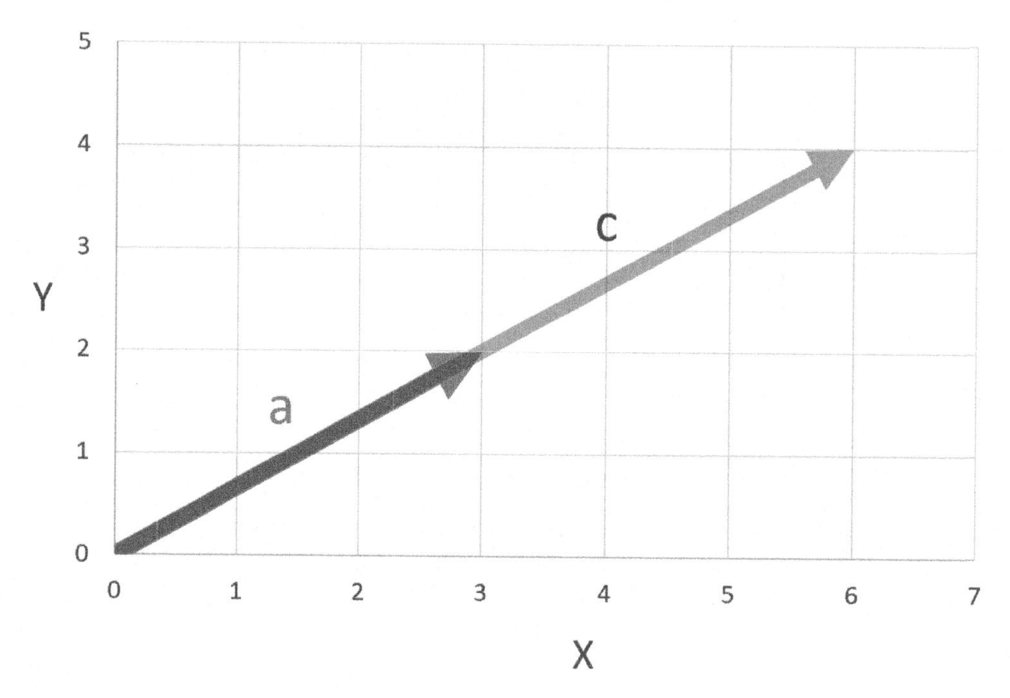

Figure 1.6 – |a⟩ scaled by two to produce |c⟩

As we have said, we can do this algebraically as well, as the following equation shows:

$$|c\rangle = 2|a\rangle = 2 \cdot \begin{bmatrix} 3 \\ 2 \end{bmatrix} = \begin{bmatrix} 6 \\ 4 \end{bmatrix}$$

Then, I want to take my vector $|b\rangle$ and scale it by two to get a new vector, $|d\rangle$, as shown in the following screenshot:

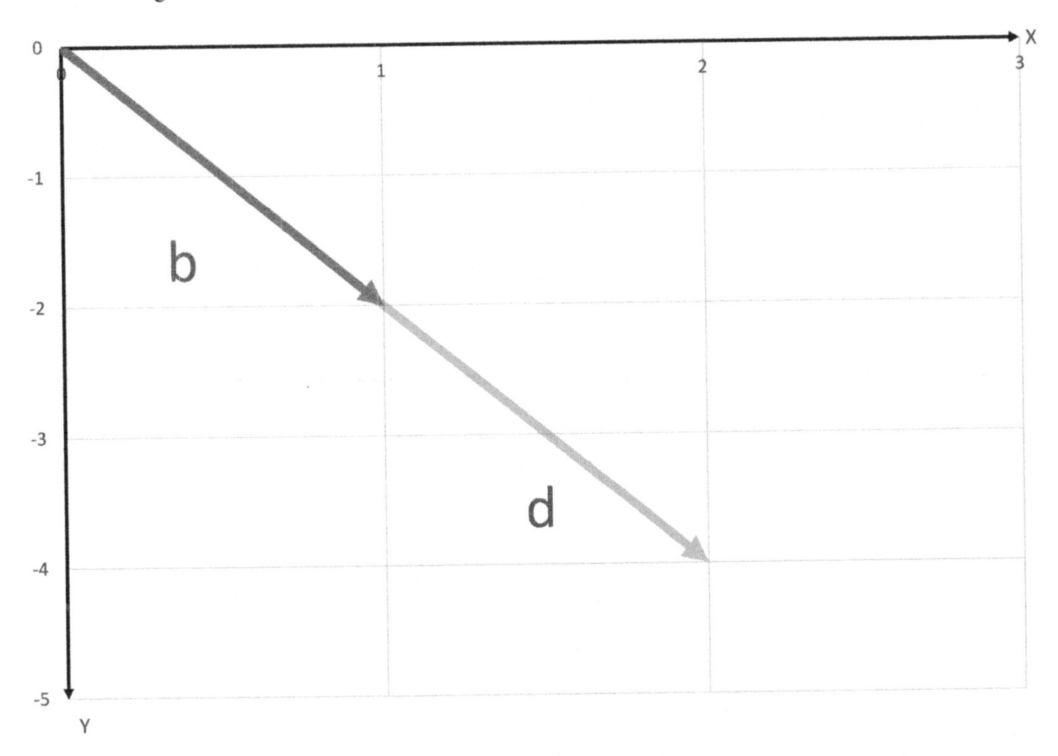

Figure 1.7 – |b⟩ scaled by two to produce |d⟩

So, now, we have a vector $|c\rangle$ that is two times $|a\rangle$, and a vector $|d\rangle$ that is two times $|b\rangle$:

$$|d\rangle = 2|b\rangle = 2\cdot\begin{bmatrix}1\\-2\end{bmatrix} = \begin{bmatrix}2\\-4\end{bmatrix}$$

Can I add these two new vectors, $|c\rangle$ and $|d\rangle$? Certainly! I will do that, but I will express $|e\rangle$ as a linear combination of $|a\rangle$ and $|b\rangle$ in the following way:

$$|e\rangle = 2|a\rangle + 2|b\rangle = |c\rangle + |d\rangle$$

Vector $|e\rangle$ is a linear combination of vectors $|a\rangle$ and $|b\rangle$! Now, I can show this all geometrically, as follows:

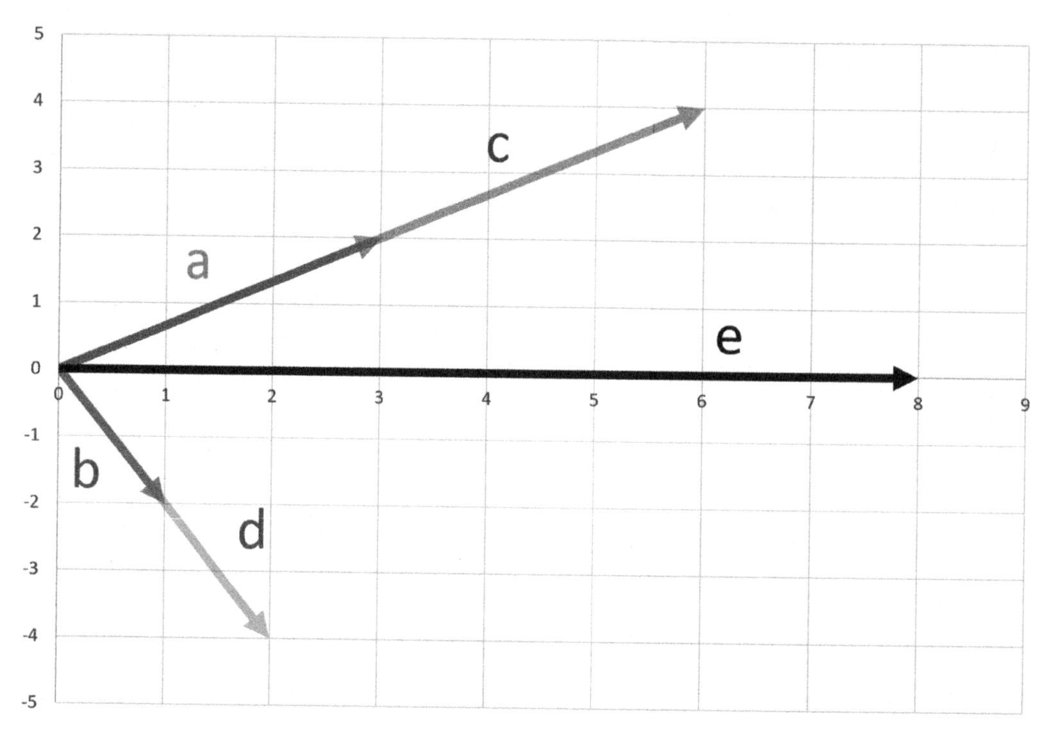

Figure 1.8 – Linear combination

This can also be represented in the following equation:

$$2\cdot\begin{bmatrix}3\\2\end{bmatrix}+2\cdot\begin{bmatrix}1\\-2\end{bmatrix}=\begin{bmatrix}6\\4\end{bmatrix}+\begin{bmatrix}2\\-4\end{bmatrix}=\begin{bmatrix}8\\0\end{bmatrix}$$

So, we now have a firm grasp on Euclidean vectors, the algebra you can perform with them, and the concept of a linear combination. We will use that in this next section to describe a quantum phenomenon called superposition.

Superposition

Superposition can be a very imposing term, so before we delve into it, let's take a step back and talk about the computers we use today. In quantum computing, we call these computers "classical computers" to distinguish them from quantum computers. Classical computers use binary digits—or **bits**, for short—to store ones and zeros. These ones and zeros can represent anything, from truth values to characters to pixel values on a screen! They are physically implemented using any two-state device such as an electrical switch that is either on or off.

A quantum bit, or **qubit** for short, is the analogous building block of quantum computers. They are implemented by anything that demonstrates quantum phenomena, which means they are very, very small. In the following screenshot, we show how a property of an electron—namely spin—can be used to represent a one or zero of a qubit:

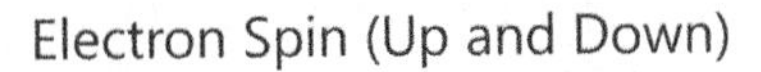

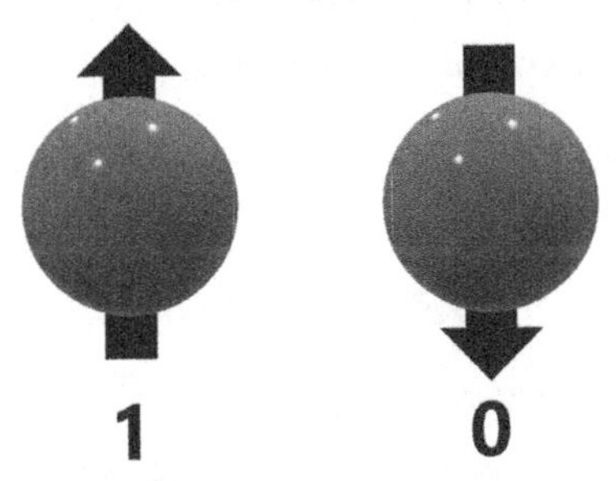

Figure 1.9 – Pair of electrons with a spin labeled 1 and 0

Physicists use mathematics to model quantum phenomena, and guess what they use to model the state of a quantum particle? That's right! Vectors! Quantum computer scientists have taken two of these states and labeled them as the canonical one and zero for qubits. They are shown in the following screenshot:

$$|1\rangle \equiv \begin{bmatrix} 0 \\ 1 \end{bmatrix}$$

$$|0\rangle \equiv \begin{bmatrix} 1 \\ 0 \end{bmatrix}$$

Y
$|1\rangle$
X
$|0\rangle$

Figure 1.10 – Zero and one states

As you can see, the zero and one states are just vectors on the x and y axes with a length of one unit each. When you combine a lot of ones and zeros in classical computing, wonderful, complex things can be done. The same is true of the zero and one state of qubits in quantum computing.

> **Greek Letters**
>
> Mathematicians and physicists love Greek letters, and they have found their way into quantum computing in several places. The Greek letter "Psi", ψ, is often used to represent the state of a qubit. The Greek letters "alpha", α, and "beta", β, are used to represent numbers or scalars.

While qubits can represent a one or a zero, they have a superpower in that they can represent a combination of a zero and one as well! "How?" you might ask. Well, this is where superposition comes in. Understanding it is actually quite simple from a mathematical standpoint. In fact, you already know what it is! It's just a fancy way of saying that a qubit is in a linear combination of states.

If you recall, we defined the vector $|e\rangle$ as a linear combination of the aforementioned $|a\rangle$ and $|b\rangle$, like so:

$$|e\rangle = 2\,|a\rangle + 2\,|b\rangle$$

Figure 1.11 – Definition of |e⟩

If we replace those letters and numbers with the Greek letters and the zero and one states we just introduced, we get an equation like this:

$$|e\rangle = 2\,|a\rangle + 2\,|b\rangle$$

$$\downarrow \quad \downarrow\downarrow \quad \downarrow\downarrow$$

$$|\psi\rangle = \alpha\,|0\rangle + \beta\,|1\rangle$$

Figure 1.12 – Greek letters being transposed onto a linear combination equation

The bottom equation represents a qubit in the state $|\psi\rangle$, which is a superposition of the states zero and one! You now know what superposition is mathematically! This, by the way, is the only way that counts because math is the language of physics and, therefore, quantum computing.

Measurement

But wait—there's more! With only the simple mathematics you have acquired so far, you also get a look at the weird act of measuring qubits. The scalars α and β shown previously play a crucial role when measuring qubits. In fact, if we were to set this qubit up in the state $|\psi\rangle$ an infinite number of times, when we measured it for a zero or a one, $|\alpha|^2$ would give us the probability of getting a zero, and $|\beta|^2$ would give us the probability of getting a one. Pretty cool, eh!?!

So, here is a question. For the qubit state $|\psi\rangle$ in the following equation, what is the probability of getting a zero or a one when we measure it?

$$|\psi\rangle = \frac{1}{\sqrt{2}}|0\rangle + \frac{1}{\sqrt{2}}|1\rangle$$

Well, if we said $|\alpha|^2$ gives us the probability of getting a zero, then the answer would look like this:

$$\left|\frac{1}{\sqrt{2}}\right|^2 = \frac{1}{\sqrt{2}} \cdot \frac{1}{\sqrt{2}} = \frac{1}{2}$$

This tells us that one half or 50% of the time when we measure for a zero or a one, we will get a zero. We can do the same exact math for β and derive that the other half of the time when we measure, we will get a one. The state $|\psi\rangle$ shown previously represents the proverbial coin being flipped into the air and landing heads for a one and tails for a zero.

Summary

In a short amount of time, we have developed enough mathematics to explain superposition and its effects on measurement. We did this by introducing Euclidean vectors and the operations of addition and scalar multiplication upon them. Putting these operations together, we were able to get a definition for a linear combination and then apply that definition to what is termed superposition. In the end, we could use all of this to predict the probability of getting a zero or one when measuring a qubit.

In the next chapter, we will introduce the concept of a matrix and use it to manipulate qubits!

History (Optional)

Euclidean vectors are named after the Greek mathematician Euclid circa 300 BC. In his book, *The Elements*, he puts together postulates and theories from other Greek mathematicians, including Pythagoras, that defined Euclidean geometry. The book was a required textbook for math students for over 2,000 years.

Figure 1.13 – Euclid with other Greek mathematicians in Raphael's School of Athens

Answers to exercises

Exercise 1

a) $\begin{bmatrix} 3 \\ 5 \end{bmatrix}$

b) $\begin{bmatrix} -3 \\ -5 \end{bmatrix}$

c) $\begin{bmatrix} 8 \\ 1 \\ -2 \end{bmatrix}$

Exercise 2

a) $\begin{bmatrix} -16 \\ 8 \end{bmatrix}$

b) $\begin{bmatrix} 8 \\ -4 \end{bmatrix}$

2
The Matrix

In the famous movie, *The Matrix*, Morpheus says, "*The Matrix is everywhere. It is all around us. Even now, in this very room.*" Morpheus was not far off the mark, for there is a theory in physics called the *matrix string theory* where essentially, reality is governed by a set of matrices. But while a matrix is one entity in the movie, a matrix in physics is a concept that is used again and again to model reality.

Figure 2.1 – This screenshot of the GLmatrix program by Jamie Zawinski is licensed with his permission

As you will see, the definition of a matrix is deceptively simple, but its power derives from all the ways mathematicians have defined that it can be used. It is a central object in quantum mechanics and, hence, quantum computing. Indeed, if I were forced to select the most important mathematical tool in quantum computing, it would be a matrix.

In this chapter, we are going to cover the following main topics:

- Defining a matrix
- Simple matrix operations
- Defining matrix multiplication
- Special types of matrices
- Quantum gates

Defining a matrix

Mathematicians define a matrix as simply a rectangular array that has *m* rows and *n* columns, like the one shown in the following screenshot:

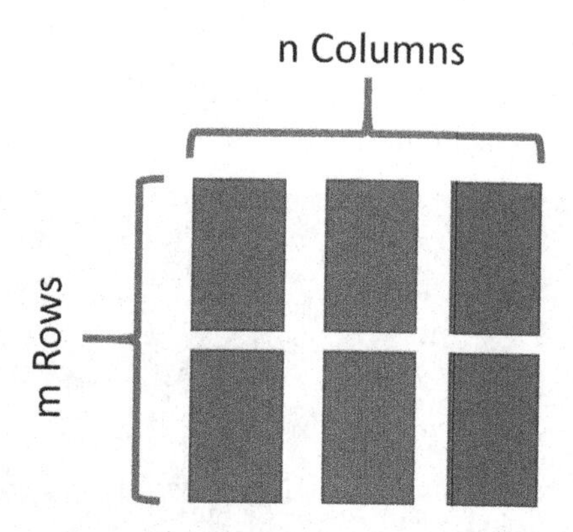

Figure 2.2 – Model of a matrix with m rows and n columns

In math, matrices are written out a particular way. An example 4×5 matrix is shown in the following expression. Notice that it has four rows and five columns:

$$\text{4 Rows} \downarrow \overset{\text{5 Columns} \rightarrow}{\begin{bmatrix} 1 & -6.2 & 4 & 5 & 7 \\ 3/2 & 22 & \sqrt{2} & 7 & -7 \\ -5 & -3.4 & 7 & 3 & 0 \\ -2 & -6 & 1.1 & -6 & 10 \end{bmatrix}}$$

Figure 2.3 – Example of a 4 x 5 matrix

Notation

In math and quantum computing, matrix variable names are in capital letters, and each entry in a matrix is referred to by a lowercase letter that corresponds to the variable name with subscripts (a_{ij}). Subscript i refers to the row the entry is in and subscript j refers to the column it is in. The following formula shows this for a 3 × 3 matrix:

$$A = \begin{bmatrix} a_{11} & a_{12} & a_{13} \\ a_{21} & a_{22} & a_{23} \\ a_{31} & a_{32} & a_{33} \end{bmatrix} = \begin{bmatrix} 5 & 4 & 3 \\ 2 & 1 & 7 \\ 8 & 9 & 0 \end{bmatrix}$$

In our example matrix A, $a_{22} = 1$. What is a_{32}? Hint—it's the only number that begins with the letter n.

Redefining vectors

One thing we will do in this book is iteratively define things so that we start simple, and as we learn more, we will add or even redefine objects to make them more advanced. So, in our previous chapter, our Euclidean vectors only had two dimensions. What if they have more? Well, we can represent them with n × 1 matrices, like so:

$$\begin{bmatrix} a_1 \\ a_2 \\ \vdots \\ a_n \end{bmatrix}$$

What do we call a 1 × n matrix such as the following one?

$$\begin{bmatrix} a_1 & a_2 & \cdots & a_n \end{bmatrix}$$

Well, we will call it a **row vector**. To distinguish between the two types, we will call the n × 1 matrix a **column vector**. Also, while we have been using kets (for example, $|x\rangle$) to notate column vectors so far, we will use something different to notate row vectors. We will introduce the bra, which is the other side of the bra-ket notation. A bra is denoted by an opening angle bracket, the name of the vector, and a pipe or vertical bar. For example, a row vector with the name b would be denoted like this: $\langle b|$. This is the other side of the bra-ket notation explained further in the appendix. To make things clearer, here are our definitions of a column vector and a row vector for now:

$$\text{column vector} := |x\rangle = \begin{bmatrix} a_1 \\ a_2 \\ \vdots \\ a_n \end{bmatrix}$$

$$\text{row vector} := \langle y| = \begin{bmatrix} a_1 & a_2 & \cdots & a_n \end{bmatrix}$$

> **Important Note**
> A bra has a deeper definition that we will look at later in this book. For now, this is enough for us to tackle matrix multiplication.

Now that we have introduced how column and row vectors can be represented by **one-dimensional** (**1D**) matrices, let's next look at some operations we can do on matrices.

Simple matrix operations

As mentioned in the introduction to this chapter, the power of matrices is the operations defined on them. Here, we go through some of the basic operations for matrices that we will build on as the book progresses. You have already encountered some of these operations with vectors in the previous chapter, but we will now expand them to matrices.

Addition

Addition is one of the easiest operations, along with its inverse subtraction. You basically just perform addition on each entry of one matrix that corresponds with another entry in the other matrix, as shown in the following formula. Addition is only defined for matrices with the same dimensions:

$$A + B = \begin{bmatrix} a_{11} & a_{12} & a_{13} \\ a_{21} & a_{22} & a_{23} \\ a_{31} & a_{32} & a_{33} \end{bmatrix} + \begin{bmatrix} b_{11} & b_{12} & b_{13} \\ b_{21} & b_{22} & b_{23} \\ b_{31} & b_{32} & b_{33} \end{bmatrix} = \begin{bmatrix} a_{11} + b_{11} & a_{12} + b_{12} & a_{13} + b_{13} \\ a_{21} + b_{21} & a_{22} + b_{22} & a_{23} + b_{23} \\ a_{31} + b_{31} & a_{32} + b_{32} & a_{33} + b_{33} \end{bmatrix}$$

Example

Here is an example of matrix addition:

$$\begin{bmatrix} 2 & 3 \\ 4 & 5 \end{bmatrix} + \begin{bmatrix} 5 & 3 \\ 0 & -1 \end{bmatrix} = \begin{bmatrix} 2+5 & 3+3 \\ 4+0 & 5-1 \end{bmatrix} = \begin{bmatrix} 7 & 6 \\ 4 & 4 \end{bmatrix}$$

Exercise 1

What is the sum of the following two matrices? (Answers to exercises are at the end of the chapter.)

$$\begin{bmatrix} 1 & 2 \\ 3 & 4 \end{bmatrix} + \begin{bmatrix} 2 & 1 \\ 0 & 2 \end{bmatrix}$$

Scalar multiplication

Scalar multiplication is also rather easy. A scalar is just a number, and so scalar multiplication is just multiplying a matrix by a number. We define it for a scalar b and matrix A thusly:

$$b \cdot A = bA = b \cdot \begin{bmatrix} a_{11} & a_{12} & \cdots & a_{1n} \\ a_{21} & a_{22} & \cdots & a_{2n} \\ \vdots & \vdots & \ddots & \vdots \\ a_{m1} & a_{m2} & \cdots & a_{mn} \end{bmatrix} = \begin{bmatrix} b \cdot a_{11} & b \cdot a_{12} & \cdots & b \cdot a_{1n} \\ b \cdot a_{21} & b \cdot a_{22} & \cdots & b \cdot a_{2n} \\ \vdots & \vdots & \ddots & \vdots \\ b \cdot a_{m1} & b \cdot a_{m2} & \cdots & b \cdot a_{mn} \end{bmatrix}$$

Example

As always, an example can definitely help:

$$3 \cdot \begin{bmatrix} 4 & 7 \\ 2 & -3 \end{bmatrix} = \begin{bmatrix} 3 \cdot 4 & 3 \cdot 7 \\ 3 \cdot 2 & 3 \cdot -3 \end{bmatrix} = \begin{bmatrix} 12 & 21 \\ 6 & -9 \end{bmatrix}$$

Exercise 2

Calculate the following:

$$3\begin{bmatrix} 3 & 4 \\ 2 & 1 \end{bmatrix}$$

Transposing a matrix

An important operation involving matrices and vectors is to transpose them. To transpose a matrix, you essentially convert the rows into columns and the columns into rows. It is denoted by a superscript *T*. Here is the definition:

$$\text{If } A = \begin{bmatrix} a_{11} & a_{12} & \cdots & a_{1n} \\ a_{21} & a_{22} & \cdots & a_{2n} \\ \vdots & \vdots & \ddots & \vdots \\ a_{m1} & a_{m2} & \cdots & a_{mn} \end{bmatrix}, \text{ then } A^T = \begin{bmatrix} a_{11} & a_{21} & \cdots & a_{m1} \\ a_{12} & a_{22} & \cdots & a_{m2} \\ \vdots & \vdots & \ddots & \vdots \\ a_{1n} & a_{2n} & \cdots & a_{mn} \end{bmatrix}$$

Notice how the subscripts for the diagonal entries stay the same, but the subscripts for all other entries are switched (for example, the entry at a_{12} becomes a_{21}). Also, as a consequence of this operation, the dimensions of the matrix are switched. A 2 × 4 matrix becomes a 4 × 2 matrix.

Examples

Here is an example of a 3 × 4 matrix transposed:

$$\begin{bmatrix} 1 & 2 & 3 & 4 \\ 5 & 6 & 7 & 8 \\ 9 & 0 & 1 & 2 \end{bmatrix}^T = \begin{bmatrix} 1 & 5 & 9 \\ 2 & 6 & 0 \\ 3 & 7 & 1 \\ 4 & 8 & 2 \end{bmatrix}$$

Here is an example of a square matrix transposed:

$$\begin{bmatrix} 3 & 5 \\ 2 & 0 \end{bmatrix}^T = \begin{bmatrix} 3 & 2 \\ 5 & 0 \end{bmatrix}$$

Now, let's move on to matrix multiplication.

Defining matrix multiplication

Matrix multiplication can be a complicated procedure, and we will build up to it gradually. It is defined as an operation between an m × n matrix and an n × p matrix that produces an m × p matrix. The following screenshot shows this well:

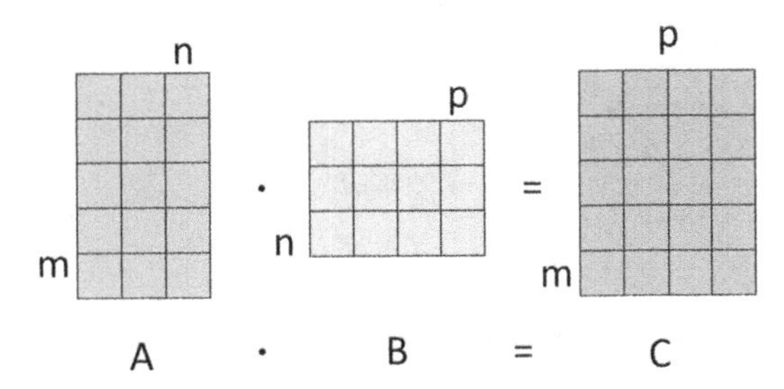

Figure 2.4 – Schematic of matrix multiplication

Notice that matrix multiplication is only defined if the number of columns in the first matrix equals the number of rows in the second matrix—or, in other words, the ns have to match in our preceding figure. This is so important that we will give it a special name: the **matrix multiplication definition rule**, or definition rule for short. Based on this, the first thing you should do when presented with two matrices to multiply is to make sure they pass the definition rule. Otherwise, the operation is undefined. For example, do the following two matrices pass the definition rule?

$$\begin{bmatrix} a & b \\ c & d \end{bmatrix} \cdot \begin{bmatrix} c & d & z \\ e & f & x \\ g & h & y \end{bmatrix}$$

The answer is no because you have a 2 × 2 matrix being multiplied by a 3 × 3 matrix. The number of columns of the first matrix *does not* equal the number of rows of the second matrix. What about the following matrix multiplication?

$$\begin{bmatrix} a & b & e \\ c & d & f \end{bmatrix} \cdot \begin{bmatrix} c & d \\ g & h \\ i & k \end{bmatrix} \tag{1}$$

Yes, it is defined! It is a 2 × 3 matrix multiplied by a 3 × 2 matrix. The number of columns in the first equals the number of rows in the second!

If the matrices pass the definition rule, the second step you should take when multiplying two matrices is to draw out the answer's dimensions. For instance, when presented with the preceding matrix multiplication from *Equation (1)* of a 2 × 3 matrix and a 3 × 2 matrix, we would draw out a 2 × 2 matrix, like so. I like to write the dimensions of each matrix on the top:

$$\overbrace{\begin{bmatrix} a & b & e \\ c & d & f \end{bmatrix}}^{2\times 3} \cdot \overbrace{\begin{bmatrix} c & d \\ g & h \\ i & k \end{bmatrix}}^{3\times 2} = \overbrace{\begin{bmatrix} \square & \square \\ \square & \square \end{bmatrix}}^{2\times 2}$$

Figure 2.5 – Writing dimensions of matrix multiplication

Remember—the number of rows in the first matrix determines the number in rows of the product or resultant matrix. The number of columns in the second matrix determines the number of columns in the product.

Okay—to summarize, here are the first two steps you should take when doing matrix multiplication:

1. Does it pass the definition rule? Do the number of columns in the first matrix equal the number of rows in the second matrix?
2. Draw out the dimensions of the product matrix. For an m × n matrix and an n × p matrix, the dimensions of the resulting matrix will be m × p.

Based on all of this, do you think matrix multiplication is commutative (that is, $A \cdot B = B \cdot A$)? Think about it—I'll give you the answer later in this chapter. Now, let's look at how to multiply two vectors to produce a scalar.

Multiplying vectors

Earlier, we defined column and row vectors as one-dimensional matrices. Since they are one-dimensional, they are the easiest matrices to multiply. A bracket, denoted by $\langle x|y\rangle$, is essentially matrix multiplication of a row vector and a column vector. Here is our definition:

$$\langle x|y\rangle = \begin{bmatrix} x_1 & x_2 & \cdots & x_n \end{bmatrix} \begin{bmatrix} y_1 \\ y_2 \\ \vdots \\ y_n \end{bmatrix} = x_1 \cdot y_1 + x_2 \cdot y_2 + \cdots + x_n \cdot y_n$$

> **Important Note**
> A bracket has a deeper definition that we will look at later in this book. For now, this is enough for us to tackle matrix multiplication.

Let's look at an example to make this more concrete.

Examples

Let's say $\langle y|$ and $|x\rangle$ are defined this way:

$$\langle y| = \begin{bmatrix} 3 & 2 & 1 & 4 \end{bmatrix} \qquad |x\rangle = \begin{bmatrix} 1 \\ 2 \\ 3 \\ 4 \end{bmatrix}$$

Now, let's calculate the bracket $\langle y| x\rangle$:

$$\langle y|x\rangle = \begin{bmatrix} 3 & 2 & 1 & 4 \end{bmatrix} \cdot \begin{bmatrix} 1 \\ 2 \\ 3 \\ 4 \end{bmatrix} = \begin{cases} 3 \cdot 1 + 2 \cdot 2 + 1 \cdot 3 + 4 \cdot 4 \\ 3 + 4 + 3 + 16 \\ 26 \end{cases}$$

Here are two more examples of matrix multiplication of a row vector with a column vector:

$$\begin{bmatrix} 3 & 2 \end{bmatrix} \begin{bmatrix} 4 \\ 1 \end{bmatrix} = 3 \cdot 4 + 2 \cdot 1 = 12 + 2 = 14$$

$$\begin{bmatrix} 1 & 2 & 3 \end{bmatrix} \begin{bmatrix} 4 \\ 5 \\ 6 \end{bmatrix} = 1 \cdot 4 + 2 \cdot 5 + 3 \cdot 6 = 4 + 10 + 18 = 32$$

Exercise 3

What is the answer to the matrix multiplication of this row vector and column vector?

$$\begin{bmatrix} 1 & 2 & 3 \end{bmatrix} \qquad \begin{bmatrix} 1 \\ 2 \\ 3 \end{bmatrix}$$

Matrix-vector multiplication

We are building up to pure matrix multiplication, and the next step to getting there is **matrix-vector multiplication**. Let's look at a typical expression for matrix-vector multiplication:

$$\begin{bmatrix} 2 & 1 \\ 3 & 4 \\ 4 & 5 \end{bmatrix} \cdot \begin{bmatrix} 2 \\ 3 \end{bmatrix}$$

Now, this is a 3×2 matrix multiplied by a 2×1 column vector. Does this pass the matrix multiplication rule? Yes! 2=2. What are the dimensions of the product? Well, taking the outer dimensions of the two matrices involved in the product, it will be a 3×1 column vector. If the matrix and column vector are variables, we can write out the product this way:

$$A|x\rangle \tag{2}$$

This denotes the matrix A multiplying the vector $|x\rangle$.

Alright—how do we actually do the multiplication? Well, first, we separate the rows of the matrix into row vectors. Wait—there are vectors in matrices?! Yes—you can see a matrix as a set of row vectors or column vectors when it is convenient to do so. Let's look at an example of doing this:

$$\begin{bmatrix} 4 & 3 & 2 \\ 4 & 1 & 3 \\ 2 & 4 & 2 \end{bmatrix} \Rightarrow \begin{matrix} \begin{bmatrix} 4 & 3 & 2 \end{bmatrix} = \langle R_1| \\ \begin{bmatrix} 4 & 1 & 3 \end{bmatrix} = \langle R_2| \\ \begin{bmatrix} 2 & 4 & 2 \end{bmatrix} = \langle R_3| \end{matrix} \tag{3}$$

See how I separated the three rows of the matrix into three row vectors? I even gave them names, with the letter R standing for row and the subscript number showing which row it came from. So, after performing the first two steps of matrix multiplication, the next step is:

3. Separate the matrix on the left into row vectors, $\langle R_1|$, $\langle R_2|$, ... , $\langle R_m|$.

From there, we will calculate the bracket between each row vector from the separated matrix and the column vector we are multiplying by. Let's see an example.

Let's say we are trying to find the answer to the matrix-vector multiplication between matrix A and vector $|x\rangle$ as in *Equation (2)* and matrix A is the one defined in *Equation (3)*. We will create a simple vector $|x\rangle$ to give us:

$$A|x\rangle = \begin{bmatrix} 4 & 3 & 2 \\ 4 & 1 & 3 \\ 2 & 4 & 2 \end{bmatrix} \cdot \begin{bmatrix} w \\ y \\ z \end{bmatrix} = \begin{bmatrix} a \\ b \\ c \end{bmatrix}$$

Okay—so, how do we figure out what a, b, and c are? Let's calculate the brackets!

$$a = \langle R_1|x\rangle = \begin{bmatrix} 4 & 3 & 2 \end{bmatrix} \begin{bmatrix} w \\ y \\ z \end{bmatrix} = 4w + 3y + 2z$$

$$b = \langle R_2|x\rangle = \begin{bmatrix} 4 & 1 & 3 \end{bmatrix} \begin{bmatrix} w \\ y \\ z \end{bmatrix} = 4w + y + 3z$$

$$c = \langle R_3|x\rangle = \begin{bmatrix} 2 & 4 & 2 \end{bmatrix} \begin{bmatrix} w \\ y \\ z \end{bmatrix} = 2w + 4y + 2z$$

Now that we have seen an example, let's generalize this for the final step of matrix-vector multiplication:

4. For a matrix-vector multiplication, $A|x\rangle$, compute the bracket for each row in A with the column vector $|x\rangle$. Put the result of the bracket in the corresponding row for the resultant column vector.

So, let's write this all out for our example:

$$A = \begin{bmatrix} 4 & 3 & 2 \\ 4 & 1 & 3 \\ 2 & 4 & 2 \end{bmatrix} \Rightarrow \begin{matrix} \begin{bmatrix} 4 & 3 & 2 \end{bmatrix} = \langle R_1| \\ \begin{bmatrix} 4 & 1 & 3 \end{bmatrix} = \langle R_2| \\ \begin{bmatrix} 2 & 4 & 2 \end{bmatrix} = \langle R_3| \end{matrix} \qquad |x\rangle = \begin{bmatrix} w \\ y \\ z \end{bmatrix}$$

$$A|x\rangle = \begin{bmatrix} 4 & 3 & 2 \\ 4 & 1 & 3 \\ 2 & 4 & 2 \end{bmatrix} \cdot \begin{bmatrix} w \\ y \\ z \end{bmatrix} = \begin{bmatrix} \langle R_1|x\rangle \\ \langle R_2|x\rangle \\ \langle R_3|x\rangle \end{bmatrix}$$

Now, let's put this all together for a proper definition of matrix-vector multiplication.

Matrix-vector multiplication definition

Given an m × n matrix A and an n × 1 vector $|x\rangle$, where A is made up of m row vectors, like so:

$$A = \begin{bmatrix} \langle R_1| \\ \langle R_2| \\ \vdots \\ \langle R_m| \end{bmatrix}$$

Matrix-vector multiplication is defined as:

$$A|x\rangle = |y\rangle = \begin{bmatrix} \langle R_1|x\rangle \\ \langle R_2|x\rangle \\ \vdots \\ \langle R_m|x\rangle \end{bmatrix}$$

Now, let's apply this to an example with real numbers:

$$\begin{bmatrix} 2 & 1 \\ 3 & 4 \\ 4 & 5 \end{bmatrix} \cdot \begin{bmatrix} 2 \\ 3 \end{bmatrix} = \begin{bmatrix} \begin{bmatrix} 2 & 1 \end{bmatrix}\begin{bmatrix} 2 \\ 3 \end{bmatrix} \\ \begin{bmatrix} 3 & 4 \end{bmatrix}\begin{bmatrix} 2 \\ 3 \end{bmatrix} \\ \begin{bmatrix} 4 & 5 \end{bmatrix}\begin{bmatrix} 2 \\ 3 \end{bmatrix} \end{bmatrix} = \begin{bmatrix} 2 \cdot 2 + 3 \cdot 1 \\ 3 \cdot 2 + 4 \cdot 3 \\ 4 \cdot 2 + 5 \cdot 3 \end{bmatrix} = \begin{bmatrix} 7 \\ 18 \\ 23 \end{bmatrix}$$

Now, it's your turn to do matrix-vector multiplication.

Exercise 4

If you have matrices A, B, and C defined as so:

$$A = \begin{bmatrix} 1 & 2 \\ 3 & 4 \\ 5 & 6 \end{bmatrix} \quad B = \begin{bmatrix} 1 & 2 \\ 3 & 4 \end{bmatrix} \quad C = \begin{bmatrix} 1 & 2 & 3 \\ 4 & 5 & 6 \end{bmatrix}$$

and three vectors defined as so:

$$|x\rangle = \begin{bmatrix} 0 \\ -1 \\ -2 \end{bmatrix} \qquad |y\rangle = \begin{bmatrix} 1 \\ -1 \end{bmatrix} \qquad |z\rangle = \begin{bmatrix} 2 \\ 2 \end{bmatrix}$$

what are the following matrix-vector products? If the operation is undefined, say so.

$$A|x\rangle \qquad C|x\rangle$$
$$A|y\rangle \qquad B|z\rangle$$

Matrix multiplication

Alright—we have finally arrived at matrix multiplication! Trust me, it is worth the wait because matrix multiplication is used all over quantum computing and you now have the basis to do the calculations correctly and succinctly.

Remember in the previous section that I said you could view a matrix as a set of row vectors? Well, it ends up you can view them as set of column vectors as well. Let's take our matrix from before and repurpose it to see matrices as a set of column vectors:

$$\begin{bmatrix} 4 & 3 & 2 \\ 4 & 1 & 3 \\ 2 & 4 & 2 \end{bmatrix} \Rightarrow \underset{|C_1\rangle}{\begin{bmatrix} 4 \\ 4 \\ 2 \end{bmatrix}} \underset{|C_2\rangle}{\begin{bmatrix} 3 \\ 1 \\ 4 \end{bmatrix}} \underset{|C_3\rangle}{\begin{bmatrix} 2 \\ 3 \\ 2 \end{bmatrix}}$$

So, this time, I separated the matrix into three column vectors. I gave each one a name with the letter C, standing for column, and the subscript number showing which column it came from.

We will use the first three steps we have defined so far for matrix multiplication as well and replace the fourth step from matrix-vector multiplication. Here are the first four steps of matrix multiplication:

1. Does it pass the definition rule? Do the number of columns in the first matrix equal the number of rows in the second matrix?
2. Draw out the dimensions of the product matrix. For an $m \times n$ matrix and an $n \times p$ matrix, the dimensions of the resulting matrix will be $m \times p$.

3. Separate the matrix on the **left** into row vectors, $\langle R_1|, \langle R_2|, \ldots, \langle R_m|$.
4. Separate the matrix on the **right** into column vectors, $|C_1\rangle, |C_2\rangle, \ldots, |C_p\rangle$.

Alright—we have arrived at the last step! Can you guess what it is? Well, it definitely involves brackets. Without further ado, here it is:

5. For each entry a_{ij} in the resultant matrix, compute the bracket of the ith row vector and the jth column vector, $\langle R_i|C_j\rangle$.

If matrix A is the left matrix and matrix B is the right matrix (that is, $A \cdot B$), then the following diagram is a good way to look at *Step 5* graphically:

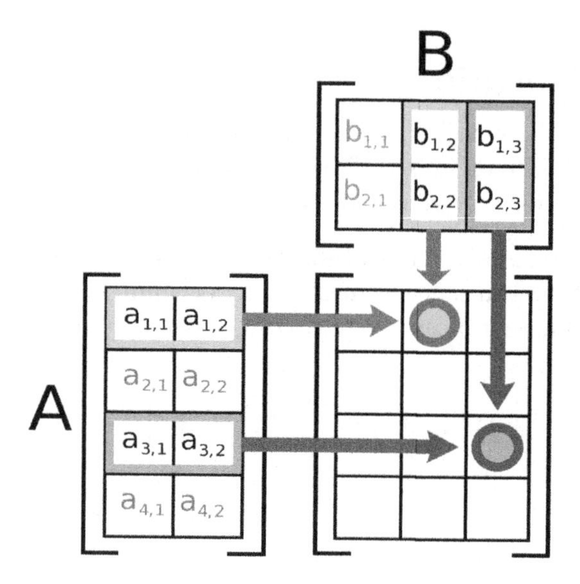

Figure 2.6 – Depiction of the matrix product AB [1]

Let's look at an example.

Example

Say we have the following two matrices:

$$A = \begin{bmatrix} 1 & 2 \\ 3 & 4 \end{bmatrix} \quad B = \begin{bmatrix} 1 & 2 & 3 \\ 4 & 5 & 6 \end{bmatrix}$$

Let's go through our five steps.

1. Does it pass the definition rule? Do the number of columns in the first matrix equal the number of rows in the second matrix?

   Yes! A is a 2×2 matrix and B is a 2×3 matrix. So, the operation is defined.

2. Draw out the dimensions of the product matrix. For an $m \times n$ matrix and an $n \times p$ matrix, the dimensions of the resulting matrix will be $m \times p$.

$$A \cdot B = \overbrace{\begin{bmatrix} 1 & 2 \\ 3 & 4 \end{bmatrix}}^{2\times 2} \cdot \overbrace{\begin{bmatrix} 1 & 2 & 3 \\ 4 & 5 & 6 \end{bmatrix}}^{2\times 3} = \overbrace{\begin{bmatrix} \square & \square & \square \\ \square & \square & \square \end{bmatrix}}^{2\times 3}$$

3. Separate the matrix on the **left** into row vectors, $\langle R_1|, \langle R_2|, \ldots, \langle R_m|$.

$$\begin{bmatrix} 1 & 2 \\ 3 & 4 \end{bmatrix} \Rightarrow \begin{matrix} \begin{bmatrix} 1 & 2 \end{bmatrix} = \langle R_1| \\ \begin{bmatrix} 3 & 4 \end{bmatrix} = \langle R_2| \end{matrix}$$

4. Separate the matrix on the **right** into column vectors, $|C_1\rangle, |C_2\rangle, \ldots, |C_p\rangle$.

$$\begin{bmatrix} 1 & 2 & 3 \\ 4 & 5 & 6 \end{bmatrix} \Rightarrow \underset{|C_1\rangle}{\begin{bmatrix} 1 \\ 4 \end{bmatrix}} \underset{|C_2\rangle}{\begin{bmatrix} 2 \\ 5 \end{bmatrix}} \underset{|C_3\rangle}{\begin{bmatrix} 3 \\ 6 \end{bmatrix}}$$

5. For each entry a_{ij} in the resultant matrix, compute the bracket of the ith row vector and the jth column vector, $\langle R_i|C_j\rangle$.

$$A \cdot B = \begin{bmatrix} \langle R_1|C_1\rangle & \langle R_1|C_2\rangle & \langle R_1|C_3\rangle \\ \langle R_2|C_1\rangle & \langle R_2|C_2\rangle & \langle R_2|C_3\rangle \end{bmatrix}$$

So, the computations will look like this:

$$\begin{bmatrix}1 & 2\\3 & 4\end{bmatrix}\cdot\begin{bmatrix}1 & 2 & 3\\4 & 5 & 6\end{bmatrix}=\begin{bmatrix}[1\ \ 2]\begin{bmatrix}1\\4\end{bmatrix} & [1\ \ 2]\begin{bmatrix}2\\5\end{bmatrix} & [1\ \ 2]\begin{bmatrix}3\\6\end{bmatrix}\\ [3\ \ 4]\begin{bmatrix}1\\4\end{bmatrix} & [3\ \ 4]\begin{bmatrix}2\\5\end{bmatrix} & [3\ \ 4]\begin{bmatrix}3\\6\end{bmatrix}\end{bmatrix}$$

$$=\begin{bmatrix}1+8 & 2+10 & 3+12\\3+16 & 6+20 & 9+24\end{bmatrix}=\begin{bmatrix}9 & 12 & 15\\19 & 26 & 33\end{bmatrix}$$

And there's our answer! That may have taken a while, but it will become quicker and more intuitive as you practice it. Speaking of practice...

Exercise 5

Given the following three matrices:

$$D=\begin{bmatrix}-1 & 2\\3 & 0\\5 & 2\end{bmatrix}\qquad E=\begin{bmatrix}1 & 2\\0 & -2\end{bmatrix}\qquad F=\begin{bmatrix}1 & 0 & 4\\-1 & 5 & -2\end{bmatrix}$$

what are the following products? Follow the steps! And if the operation is undefined, say so.

$$D\cdot E$$
$$E\cdot F$$
$$F\cdot D$$

Properties of matrix multiplication

It's good to know some general properties of matrix multiplication. These properties assume that the matrices all have the right dimensions to pass the matrix multiplication definition rule. Here they are—matrix multiplication is:

- Not commutative: $A\cdot B \neq B\cdot A$
- Distributive with respect to matrix addition:

$$A(B+C) = AB + AC = (B+C)A$$

- Associative:

$$(AB)C = A(BC)$$

- The transpose of a matrix product $A \cdot B$ is the product of the transpose of each matrix in reverse:

$$(AB)^T = B^T A^T$$

That concludes our section on matrix multiplication. You might want to take a break—you've been through a lot! When you come back, we'll look at some special matrices that are good to know.

Special types of matrices

In the world of matrices, some are so special that they have been singled out. Here they are.

Square matrices

A special type of matrix is a **square matrix**. A square matrix is one where the number of rows equals the number of columns. In other words, it is an m × n matrix in which m = n. Square matrices show up all over the place in quantum computing due to special properties that they can have—for example, symmetry, which is discussed later in the book. As we progress in the book, they will become one of the central types of matrices we will use. Some examples of square matrices are:

$$\begin{bmatrix} 2 & 5 \\ 1 & 7 \end{bmatrix} \quad \begin{bmatrix} 3 & 8 & 8 \\ 9 & 1 & 6 \\ 3 & 5 & 6 \end{bmatrix} \quad \begin{bmatrix} 3 & 8 & 4 & 6 \\ 2 & 9 & 2 & 9 \\ 1 & 6 & 0 & 5 \\ 0 & 7 & 5 & 4 \end{bmatrix}$$

Identity matrices

An important type of square matrix is an **identity matrix**, named I. It is defined so that it acts as the number 1 in matrix multiplication so that the following holds true:

$$A \cdot I = A$$

It has ones all down its principal diagonal and zeros everywhere else. Its dimensions need to change based on the matrix it is being multiplied by. Here are some examples of I in different dimensions:

$$I_1 = [1]\ ,\ I_2 = \begin{bmatrix} 1 & 0 \\ 0 & 1 \end{bmatrix},\ I_3 = \begin{bmatrix} 1 & 0 & 0 \\ 0 & 1 & 0 \\ 0 & 0 & 1 \end{bmatrix}$$

You should multiply some of the matrices we have used before with the identity matrix to convince yourself that it does indeed return the matrix it is multiplied by. Now that we've gone over some special matrices, let's get into why matrix multiplication is important in quantum computing.

Quantum gates

In this section, I'd like to take the math you have learned in this chapter around matrices and connect it to actual quantum computing—namely, quantum gates. Please remember that this book is not about teaching you everything in quantum computing, but rather the mathematics needed to do and learn quantum computing. That being said, I want to connect the math to quantum computing and show the motivation for learning it. Do not be frustrated if this does not all make sense, and please consult the reference books in the appendix for more information on quantum gates.

Logic gates

In classical computing, we use **logic gates** to put together circuits that will implement algorithms, such as, adding two numbers. The logic gates represent Boolean logic. Here are some simple logic operations:

- `AND`
- `OR`
- `NOT`

In a circuit, you have input, output, and logic gates. The input and outputs are represented by a binary number, with `1` being true and `0` being false. Here is an example of a `NOT` gate:

Input Output

Figure 2.7 – NOT gate

Truth tables can be created that correspond to the inputs and outputs of the circuit. Here is a truth table for the `NOT` gate just described:

Input	Output
1	0
0	1

Figure 2.8 – Truth table for NOT gate

A slightly more complicated circuit involves the `AND` gate, as shown in the following figure:

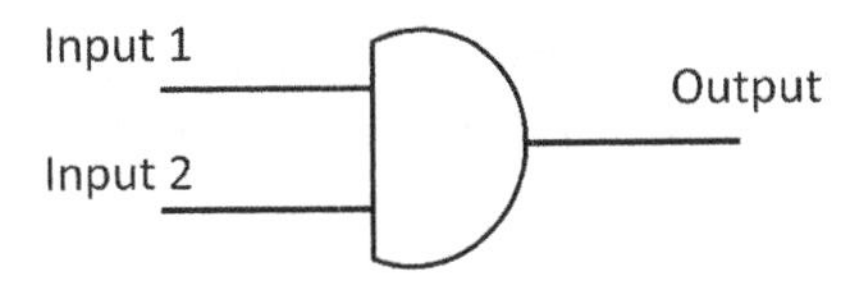

Figure 2.9 – AND gate

Here is a truth table for this circuit:

Input 1	Input 2	Output
0	0	0
0	1	0
1	0	0
1	1	1

Figure 2.10 – Truth table for AND gate

Circuit model

Much of quantum computing is modeled using quantum circuits that are similar to, but not the same as, the classical circuits we just went through. In quantum circuits, the inputs are qubits (vectors), and the gates are matrices. An example quantum logic gate is shown here:

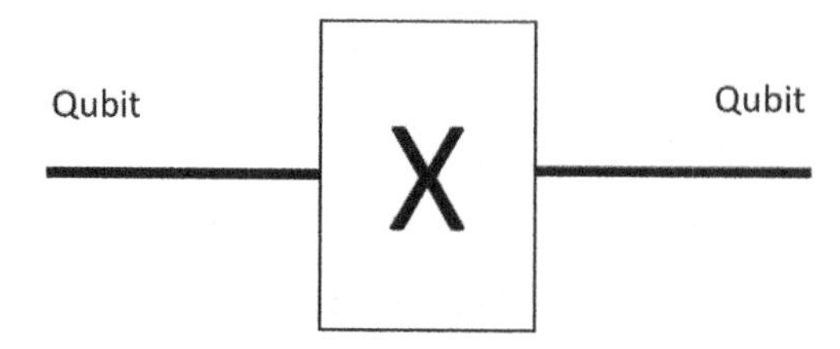

Figure 2.11 – Quantum circuit with NOT gate

The output qubit is derived through matrix-vector multiplication! So, if we remember from *Chapter 2, Superposition with Euclid*, qubits are just vectors and the binary states for qubits are:

$$|0\rangle = \begin{bmatrix} 1 \\ 0 \end{bmatrix} \text{ and } |1\rangle = \begin{bmatrix} 0 \\ 1 \end{bmatrix}$$

The `NOT` gate in quantum computing is represented by the following matrix:

$$X = \begin{bmatrix} 0 & 1 \\ 1 & 0 \end{bmatrix}$$

So, if our input qubit is a $|1\rangle$, then the output would be:

$$X|1\rangle = \begin{bmatrix} 0 & 1 \\ 1 & 0 \end{bmatrix}\begin{bmatrix} 0 \\ 1 \end{bmatrix} = \begin{bmatrix} 1 \\ 0 \end{bmatrix} = |0\rangle$$

If we did the computation with $|0\rangle$ as the input, the output would be $|1\rangle$. Basically, this is the quantum version of the `NOT` gate!

There are many quantum gates, but they are all modeled as matrices. Thus, the math you have learned in this chapter is directly applicable to quantum computing!

Summary

In this chapter, we have learned a good number of operations on matrices and vectors. We can now do basic computation with them, and we saw their application to quantum computing. Please note that not all matrices can be used as quantum gates. Please keep reading the book to find out which ones can.

In the next chapter, we start to go deeper and look at the foundations of mathematics.

Answers to exercises

Exercise 1

$$\begin{bmatrix} 3 & 3 \\ 3 & 6 \end{bmatrix}$$

Exercise 2

$$\begin{bmatrix} 9 & 12 \\ 6 & 3 \end{bmatrix}$$

Exercise 3

14

Exercise 4

$$A|x\rangle \text{ is undefined} \qquad C|x\rangle = \begin{bmatrix} -8 \\ -17 \end{bmatrix}$$

$$A|y\rangle = \begin{bmatrix} -1 \\ -1 \\ -1 \end{bmatrix} \qquad B|z\rangle = \begin{bmatrix} 6 \\ 14 \end{bmatrix}$$

Exercise 5

$$D = \begin{bmatrix} -1 & 2 \\ 3 & 0 \\ 5 & 2 \end{bmatrix} \quad E = \begin{bmatrix} 1 & 2 \\ 0 & -2 \end{bmatrix} \quad F = \begin{bmatrix} 1 & 0 & 4 \\ -1 & 5 & -2 \end{bmatrix}$$

$$D \cdot E = \begin{bmatrix} -1 & 2 \\ 3 & 0 \\ 5 & 2 \end{bmatrix} \begin{bmatrix} 1 & 2 \\ 0 & -2 \end{bmatrix} = \begin{bmatrix} -1 & -6 \\ 3 & 6 \\ 5 & 6 \end{bmatrix}$$

$$E \cdot F = \begin{bmatrix} 1 & 2 \\ 0 & -2 \end{bmatrix} \begin{bmatrix} 1 & 0 & 4 \\ -1 & 5 & -2 \end{bmatrix} = \begin{bmatrix} -1 & 10 & 0 \\ 2 & -10 & 4 \end{bmatrix}$$

$$F \cdot D = \begin{bmatrix} 1 & 0 & 4 \\ -1 & 5 & -2 \end{bmatrix} \begin{bmatrix} -1 & 2 \\ 3 & 0 \\ 5 & 2 \end{bmatrix} = \begin{bmatrix} 19 & 10 \\ 6 & -6 \end{bmatrix}$$

References

[1] *Matrix multiplication diagram* by *Bilou* is licensed under *CC BY-SA 3.0.*

Section 2: Elementary Linear Algebra

This section digs deeper into the heart of quantum computing: linear algebra.

The following chapters are included in this section:

- *Chapter 3, Foundations*
- *Chapter 4, Vector Spaces*
- *Chapter 5, Using Matrices to Transform Space*

3
Foundations

Up until this point, we have introduced our mathematics with as little rigor as possible. This chapter – and this part of the book – will change that. You may ask why? Well, this rigor and foundational material is needed when we get to much more complex concepts such as **Hilbert spaces** and **tensor products**. Without this chapter, these advanced concepts will not make sense, and you won't have the context to understand them.

This chapter goes through the field of **abstract algebra**. As you might expect, there will be some abstract concepts that will be explored. Abstract algebra takes a step back from all other forms of algebra, such as linear, Boolean, and elementary algebra, and it tries to see what can be generalized between them. Mathematicians have found that they can generalize a few foundational concepts that, when put together, allow us to go further in math than we have before and help us understand it at a more fundamental level. Within this chapter, our ultimate goal will be to define **vector spaces** rigorously.

So, without further ado, let's get into the material. We will cover the following topics:

- Sets
- Functions
- Binary operations
- Groups
- Fields
- Vector spaces

Sets

Sets are very intuitive and are really about grouping things together. For example, all mammals – taken together – form a set. In this set, its members are things such as the fox, squirrel, and dog. Sets don't care about duplication – so, if we have 5,000 dogs in our first mammals set, this set is equal to a set of mammals that has only one dog. Let's make this more formal.

The definition of a set

A *set* is a *collection of objects*. This collection can be finite or infinite. Mathematical objects are abstract, have properties, and can be acted upon by operations. Examples of objects are numbers, functions, shapes, and matrices. Objects in a set are called **elements** or **members**.

Notation

There are multiple ways to denote a set. The easiest way is to just describe it, as I did with the set of mammals. Another example for doing this would be to describe a set S of all US States. Some examples of elements in this set would be Virginia and Alabama. Let's look at a more formal way to notate sets called **set-builder notation.**

Set-builder notation

Set-builder notation is definitely more formal, but with this formality, you gain *preciseness* (which mathematicians covet). The easiest way to denote a set in set-builder notation is just to enumerate all the members of a set. You do this by giving the variable name of the set, followed by an equals sign, and then the members of the set are put into curly brackets and separated by commas. Here is an example:

$$N = \{He, Ne, Ar, Kr, Xe, Rn, Og\}$$

Any guesses on what this set is? Extra bonus points if you guessed the noble gases.

An ellipsis (...) is used to skip listing elements if a pattern is clear or to denote an infinite set, shown as follows:

$$X = \{1,\ 2,\ 3,\ \ldots,\ 100\} \quad Y = \{\ldots,\ -2,\ -1, 0, 1, 2, \ldots\}$$

The final way you can denote a set is to include conditions for the members of your set. Here is an annotated example to explain each part of the notation:

$$\{\, x \mid x < 10 \,\}$$

The set of — all x — such that — Some condition

Figure 3.1 – An annotated description of set-builder notation

You can also describe what type of number you are dealing with by writing the type before the vertical bar, like so:

$$\{x \text{ is a prime number} | x < 8\}$$

This is equivalent to the set {2, 3, 5, 7}.

Other set notation

An important symbol in set notation is ϵ, which denotes membership. If there is a slash through ϵ, it means the object is not a member of the set. For example, the following denotes that *He* (helium) is a member of the noble gases and *O* (oxygen) is not:

$$He \in N \qquad O \notin N$$

The next symbol to consider is used to denote subsets. If X and Y are sets, and every element of X is also an element of Y, then:

- X is a **subset** of Y, denoted by $X \subseteq Y$.
- Y is a **superset** of X, denoted by $Y \supseteq X$.

For example, the set N of the noble gases is a subset of the set E of all elements. Or, mammals are a subset of the animal kingdom and a superset of the primates.

Finally, it is important to define the empty set that has no members and is denoted by $\emptyset$ or {}. The empty set is a subset of all sets.

Important sets of numbers

Since mathematics is all about numbers, special attention should be given to certain sets of numbers that we will see in this book. Each one has a special double-struck capital letter to represent them. So, without further ado, here is the list:

- $\mathbb{N}$, which is the set of natural numbers defined as {0, 1, 2, 3, ...}. This is the first set of numbers you learn as a child, and they are used for counting.

- $\mathbb{Z}$, which is the set of integers defined as {..., -3, -2, -1, 0, 1, 2, 3, ...}. This is a superset of $\mathbb{N}$. The letter *Z* comes from the German word *Zahlen*, which means *numbers*.
- $\mathbb{Q}$, which is the set of rational numbers, where a *rational number* is defined as any number that can be expressed as a ratio or quotient of two integers. Since all integers are divisible by 1, $\mathbb{Z}$ is a subset of $\mathbb{Q}$.
- $\mathbb{R}$, which is the set of real numbers. The *real numbers* are composed of $\mathbb{Q}$ and all the irrational numbers. Irrational numbers, when represented as a decimal, do not terminate, nor do they end with a repeating sequence. For example, the rational number $1/3$ is represented as $0.33333...$ in decimal, but the irrational number π starts with 3.14159, but it never terminates nor repeats a sequence. Some other examples of irrational numbers are as follows: all square roots of natural numbers that are not perfect squares; the golden ratio, ϕ; Euler's number, *e*.
- $\mathbb{C}$, which is the set of complex numbers. We will go in-depth into complex numbers in a later chapter, but for now, we will just say that a *complex number* is represented by:

$$a + bi \quad \text{where } a \text{ and } b \text{ are real numbers and } i^2 = -1$$

If we set $b = 0$, then we have the set of all real numbers, so $\mathbb{R} \subseteq \mathbb{C}$.

To sum up:

$$\mathbb{N} \subseteq \mathbb{Z} \subseteq \mathbb{Q} \subseteq \mathbb{R} \subseteq \mathbb{C}$$

The following diagram shows this very well graphically:

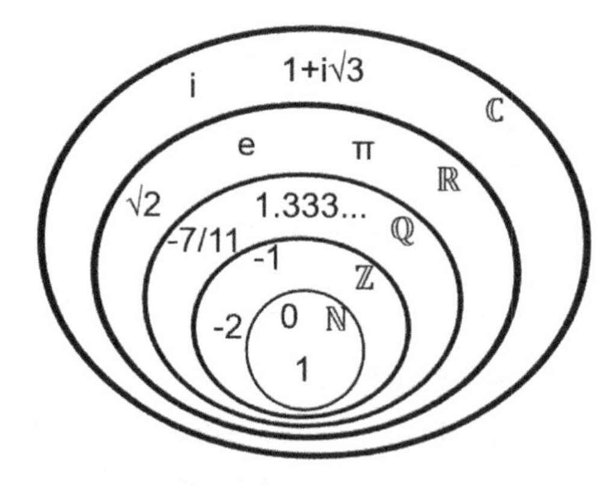

Figure 3.2 – A diagram showing all the sets of numbers [1]

Alright, let's move on to tuples!

Tuples

It is important to note that sets do not care about order. So, if set $A = \{1, 2, 3\}$ and set $B = \{2, 3, 1\}$, A and B are equal. Sets also do not care about duplication. So, if set $C = \{1, 2, 3, 3, 3\}$ and set $D = \{1, 2, 3\}$, C and D are also equal. A mathematical object that *does care* about these things is called a tuple.

A **tuple** is a finite, ordered list of elements, which is denoted with open and close parentheses, as shown here:

$$E = (3,4,5,5) \quad F = (3,4,5) \quad H = (5,3,4)$$

Since order and duplication matters to tuples, none of the preceding examples are equal to each other. The number of elements in a tuple is defined as n, and we use this number to refer to a tuple as an *n-tuple*. For example, in the preceding example, E is a 4-tuple and F is a 3-tuple. Some n-tuples have special names; for example, a 2-tuple is also known as an *ordered pair*.

The Cartesian product

The **Cartesian product** may not be as familiar to you as sets, but it is still important. The Cartesian product takes two sets and creates a third set of ordered pairs (that is, 2-tuples) from those two sets. It is denoted by the × symbol. The following figure shows an example for the Cartesian product of two sets $A=\{x, y, z\}$ and $B=\{1, 2, 3\}$.

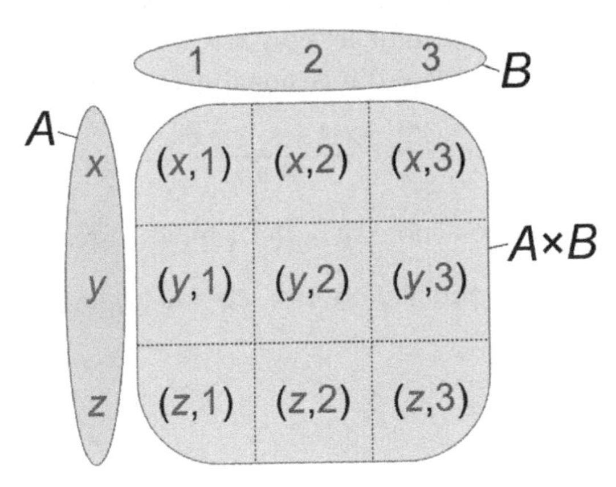

Figure 3.3 – An example of the Cartesian product of A × B [2]

Here's another example: if I have a set $A = \{1, 2\}$ and a set $B = \{6, 7, 8, 9\}$, then $A \times B$ is equal to $\{(1, 6), (1, 7), (1, 8), (1, 9), (2, 6), (2, 7), (2, 8), (2, 9)\}$. It should be noted that the Cartesian product is not commutative, so in general, $A \times B \neq B \times A$.

The greatest example of this operation is from Rene Descartes (who the Cartesian product is named after). You have probably heard of the **Cartesian plane**, as shown in the following diagram. Well, this is the Cartesian product of the set X of the real numbers $\mathbb{R}$ and the set Y of the real numbers $\mathbb{R}$, and it is denoted by $\mathbb{R} \times \mathbb{R} = \mathbb{R}^2$.

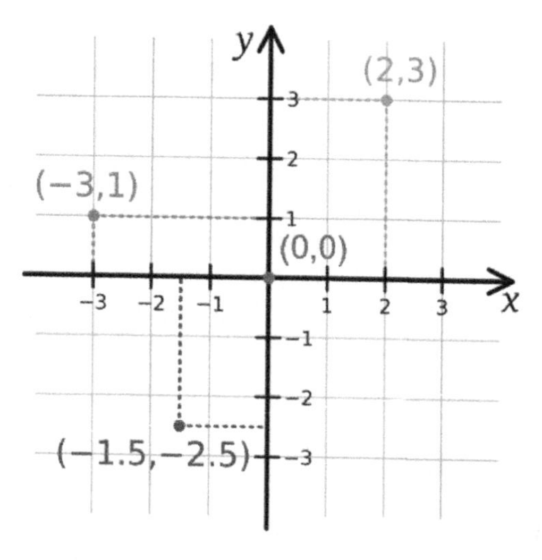

Figure 3.4 – A Cartesian plane with Cartesian coordinates [3]

The Cartesian product can be done multiple times, so if you have three sets, for example, X, Y, and Z, then $X \times Y \times Z$ is the set of all 3-tuples for every combination of elements of X, Y, and Z. Again, an example is helpful: let's say $\mathbb{R} \times \mathbb{R} \times \mathbb{R}$ produces $\mathbb{R}^3$, which is the set of all 3-tuples of real numbers or three-dimensional space. In general:

$$X^n = \underbrace{X \times X \times \ldots \times X}_{n\ times}$$

So, for shorthand, we write $\mathbb{R}^n$ to denote all of the n-tuples of real numbers.

Now that we have covered everything to do with sets and tuples, let's look at another fundamental object: functions.

Functions

Functions are fundamental to mathematics, and there is no doubt that you have been exposed to them before. However, I want to go over certain aspects of them in depth, as we will define things such as matrices as representations of functions later in the book.

The definition of a function

A **function**, for example, $y = f(x)$, maps every element x in a set A to another element y in set B. Each element y is called the **image** of x under the function $f(x)$. Set A is called the **domain** of the function and set B is called the **codomain** of the function. The domain and codomain of a function are denoted by $f: A \rightarrow B$. The following mapping diagram shows the function $f: X \rightarrow Y$.

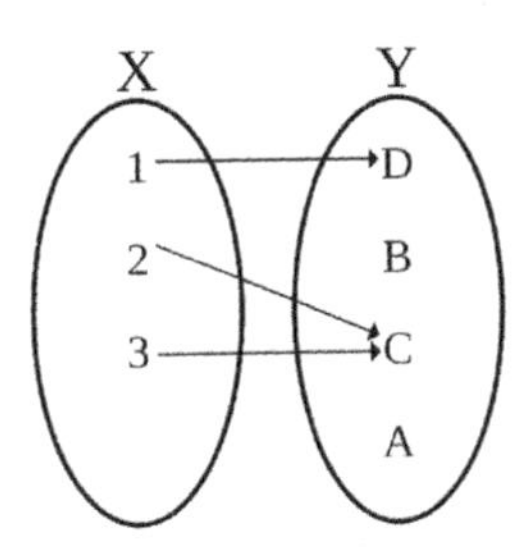

Figure 3.5 – An example function [4]

All the images of $f(x)$ form a set called the **range**. The range is a subset of the codomain. In the previous diagram of our function $f: X \rightarrow Y$, the codomain was the set Y, but the range was the set $\{D, C\}$. The image of the element 1 in the domain was the element D in the range.

I could define another function, $f: \mathbb{R} \rightarrow \mathbb{R}$, where $f(x) = 2x$. Here, the domain and codomain are the real numbers, as well as the range. The image of 3 under $f(x)$ is $f(3)=6$.

There are two rules that functions must follow:

1. Every member of the domain must be mapped.
2. Every member of the domain cannot be mapped to more than one element in the codomain.

The mapping shown in the following figure is illegal because it doesn't follow these two rules. It breaks the first rule by not mapping the elements 3 and 4 in the domain. Can you spot how it breaks the second rule?

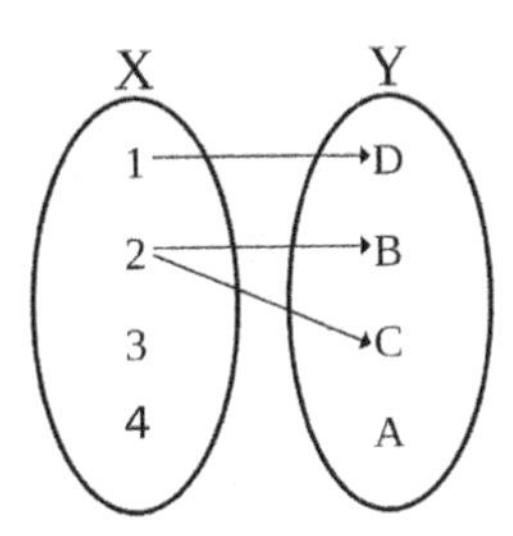

Figure 3.6 – An example of an illegal function [5]

I'm sure you pointed out that it breaks the second rule by mapping the element 2 to *B and C*.

Let's say I have a set $C = \{1, 2, 3\}$ and a set $D = \{4, 5, 6\}$. One of the many ways I can define a function is with a table. This table defines a function, $f{:}C \rightarrow D$.

x	f(x)
1	5
2	4
3	6

Figure 3.7 – A function table

Now, imagine I delete the last row from the table. Is it still a function? No, because I do not have a mapping for every element of the domain set A, namely, the number 3.

Exercise 1

For the sets $E = \{a, b, c\}$ and $F = \{4, 5, 6, 7, 8\}$, and a function, $f{:}E \rightarrow F$, which of the following tables do not represent a function?

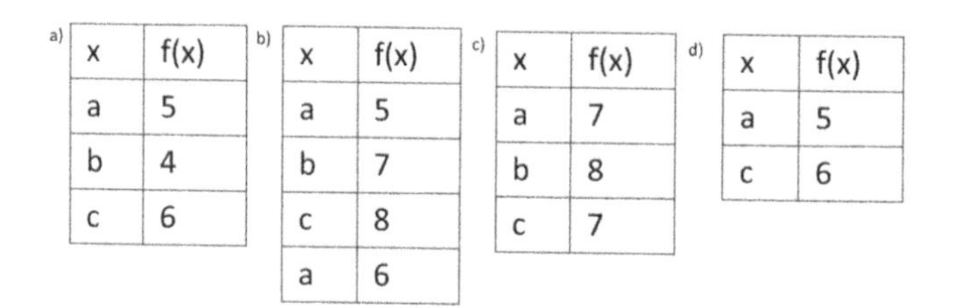

a)

x	f(x)
a	5
b	4
c	6

b)

x	f(x)
a	5
b	7
c	8
a	6

c)

x	f(x)
a	7
b	8
c	7

d)

x	f(x)
a	5
c	6

Figure 3.8 – The Exercise 1 tables

Invertible functions

Invertible functions are key in quantum computing because the laws of quantum mechanics only allow these types of functions in certain situations. Before going into invertibility, I'd like to look at three other properties of functions.

Injective functions

An **injective** function, also known as a *one-to-one* function, is a function where each element in the range is the image of only *one* element in the domain. It is important to note that not every element in the codomain needs to be in the range, so a function is still injective if there are members of the codomain that are not mapped. Let's look at an example.

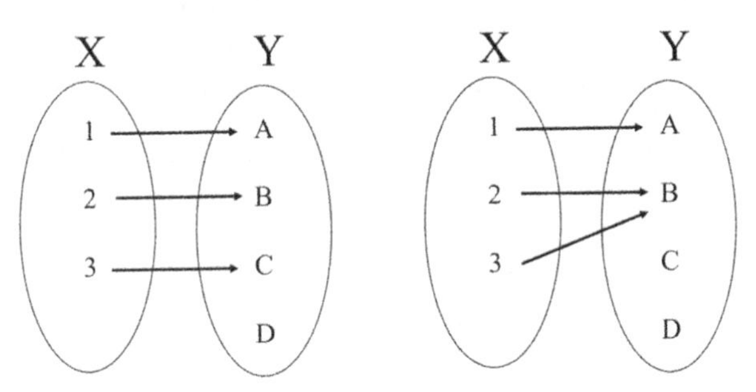

Figure 3.9 – The function on the left is injective and the function on the right is not

The function on the left in *Figure 3.9* is injective because A, B, and C are the image of only one element in the domain X. The function on the right is not injective because B is the image of both the numbers 2 and 3 in the domain X.

Surjective functions

A **surjective** function, also known as an *onto* function, is a function where the range of the domain of the function is equal to its codomain. Another way to say this is that every element in the codomain is mapped to by at least one element in the domain.

Neither of the functions from *Figure 3.9* is surjective because they leave elements in the codomain unmapped. However, the following functions are surjective:

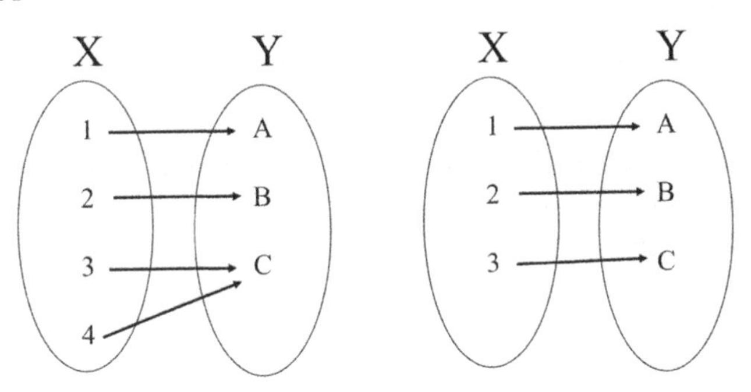

Figure 3.10 – Two surjective functions

The function $f: \mathbb{R} \rightarrow \mathbb{R}$, where $f(x) = x^2$, is not surjective because not all of the elements in the codomain (namely, all negative real numbers) are mapped to, as the square of a number (that is, x^2) can only be non-negative. The negative real numbers are left out.

Bijective functions

Now that we know what injective and surjective functions are, defining a **bijective** function is quite easy! A bijective function is one that is both injective *and* surjective. And guess what else we get in this deal!? A function is invertible if it is bijective!

Now, did you notice any bijective functions in our preceding examples? Extra, extra bonus points if you pointed to the function on the right in *Figure 3.10*, which is reproduced in the following figure:

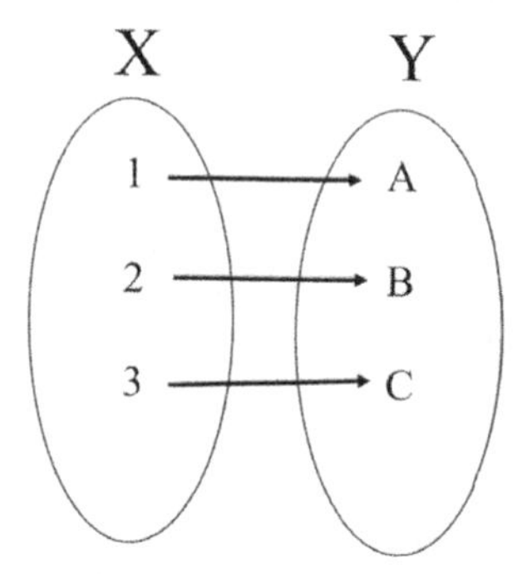

Figure 3.11 – A bijective, invertible function

The inverse of a function, such as, $f(x)$ is usually denoted by $f^{-1}(x)$. It makes sense that an invertible function has to be bijective. The only way I can make $f^{-1}(x)$ a function is to make it follow the two rules for functions that we described before, namely:

- Every member of the domain must be mapped.
- Every member of the domain cannot be mapped to more than one element in the codomain.

For $f^{-1}(x)$, the domain and codomain are flipped. For it to follow the first rule, every element of the codomain for $f(x)$ must be mapped (that is, $f(x)$ has to be surjective). And for it to follow the second rule, every member of the range of $f(x)$ cannot be mapped to more than one element in its domain (that is, $f(x)$ is injective). The following figure shows $f^{-1}(x)$ graphically:

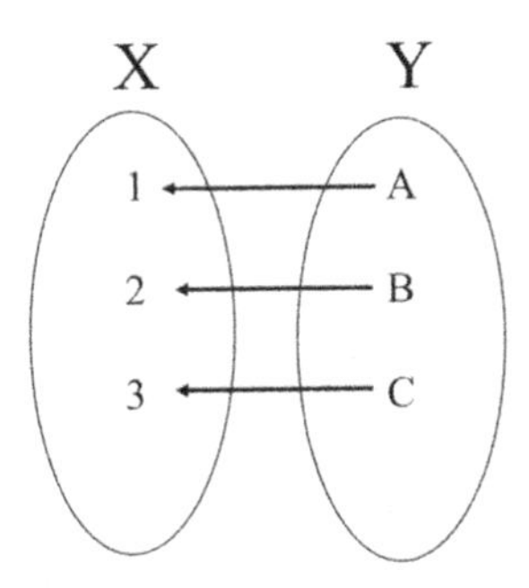

Figure 3.12 – The inverse of f(x)

Alright, that concludes our discussion of functions. Let's move on to binary operations.

Binary operations

You are probably familiar with some binary operations, for example, addition and multiplication, but we are going to look at binary operations in more depth.

The definition of a binary operation

A **binary operation** is simply a function that takes two input values and outputs one value. More precisely, it takes an ordered pair (known as an **operand**) from the Cartesian product of two sets and produces an element in another set. Using our notation:

$$f: A \times B \to C$$

An operation can be anything! For example, sexual reproduction within the set of mammals can be considered a binary operation. It takes an ordered pair from the subsets of males and females and produces another member of the set of mammals. More formally:

$$\text{sexual reproduction : male mammals} \times \text{female mammals} \to \text{mammal}$$

Within the number systems, addition is a good example of a binary operation. Let's define it for the real numbers:

$$f: \mathbb{R} \times \mathbb{R} \to \mathbb{R}$$

$$f(x, y) = x + y$$

You'll notice that with binary operations, we don't use the usual function notation of *f(x, y)*, but instead, we use a symbol with the first element on the left side and the second element on the right side. I want to remind you that binary operations are a general concept – that is, they can be *anything* – so I will use the unusual ❁ symbol when I want to talk about operations in general.

Properties

Operations can have several properties, most of which are probably familiar to you from grade school. But again, it's important to spell these out to understand abstract algebra. Here are the properties for a set S:

1. **Identity**: There exists an identity element, $e \in S$, such that for all $x \in S$, $a \circledast e = e \circledast a = a$. This element, e, is unique, and it is called the **identity** element of the group.
2. **Associativity**: If $a, b, c \in S$, then $a \circledast (b \circledast c) = (a \circledast b) \circledast c$.
3. **Invertibility**: For every $a \in S$, there exists an a^{-1}, such that $a \circledast a^{-1} = a^{-1} \circledast a = e$, where e is the identity element identified in rule one.
4. **Closure**: For every $a, b \in A$, $a \circledast b$ produces an element c that is also in the set $A. f\colon A \times A \rightarrow A$.
5. **Commutativity**: If $a, b \in S$, then $a \circledast b = b \circledast a$.

Okay, now that we defined binary operations and their properties, we can move on to discuss important algebraic structures.

Groups

A **group** builds upon the concept of a **set** by adding a binary operation to it. We denote a group by putting the set and the operation in angle brackets ($\langle\rangle$). For example, $\langle A, \circledast\rangle$ for set A and operation $\circledast$. The operation has to follow certain rules to be considered a group, namely, the rules of **identity**, **associativity**, **invertibility**, and **closure**. If the operation $\circledast$ also has the property of **commutativity**, then it is called an **Abelian group** (also known as a **commutative group**).

In our example set of mammals, the operation of sexual reproduction would not make it a group because the only property it has is commutativity.

Now, let's look at a mathematical example. What if we define $\circledast$ to be addition over the natural numbers $\mathbb{N}$ denoted $\langle\mathbb{N}, \circledast\rangle$ – is this a group? Well, let's go through the properties and see if it fulfills each one.

1. **Identity**: Does there exist an identity element e, such that $a + e = e + a = a$? Well, yes, if we define $e = 0$!
2. **Associativity**: Does $a + (b + c) = (a + b) + c$? Yes!

3. **Invertibility**: For every $a \in \mathbb{N}$, is there an a^{-1}, such that $a + a^{-1} = a^{-1} + a = e$, where e is the identity element identified in rule one ($e = 0$)? Hmmm, this is a tough one. So, $\mathbb{N}$ starts at zero and goes to positive infinity, but it does not include the negative numbers. Without negative numbers, there is no way to define an inverse of a that when added to a, will always equal zero.
4. **Closure**: If we take two numbers, a and b, in $\mathbb{N}$, then does $a + b$ produce a natural number, c? Well, zero is taken care of because it is our identity element for rule one. $1 + 2 = 3$, and 3 is also in $\mathbb{N}$. How about 100,000 + 200,000? Well, that equals 300,000, and that is also in $\mathbb{N}$. So, no matter how large we pick two numbers in $\mathbb{N}$, they will always produce another number in $\mathbb{N}$, as it goes to positive infinity!

So, there you have it. It ends up that addition with the set $\mathbb{N}$ is not a group. Is there a set that would work? Why yes! The set of all integers, $\mathbb{Z}$! We can then define a^{-1} to be $-a$, and suddenly, invertibility is fulfilled because $a + (-a) = 0$! Therefore, the set $\mathbb{Z}$ with the operation of addition qualifies as a group! Since addition is commutative as well, the group is also an Abelian group.

Fields

Fields extend the concept of groups to include another operation. Now, mathematicians end up defining fields with the familiar symbols of $\cdot$ and +, and they even call them multiplication and addition, but hopefully by now, you can see that in abstract algebra, these are just general terms that can mean anything. So, without further ado, let's define a field.

A **field** is a set (denoted by S) and two operations (+ and $\cdot$) that we will notate as $\{S, +, \cdot\}$, which follows these rules:

1. $\langle S, +\rangle$ is an Abelian group with the identity element $e = 0$.
2. If you exclude the number 0 from the set S to produce a new set S', then $\langle S', \cdot\rangle$ is an Abelian group with the identity element $e = 1$.
3. For the rule of distributivity, let $a, b, c \in S$. Then, $a \cdot (b + c) = a \cdot b + a \cdot c$.

The set of real numbers $\mathbb{R}$, with the operations of addition and multiplication, is the most obvious example of a field, but there are plenty of others.

Exercise 2

Is the set $\mathbb{R}$, with the operations of subtraction and division $\{\mathbb{R}, -, \div\}$, a field?

Vector space

Now that we have covered all of the abstract concepts we need to understand, we can give a formal definition of a vector space, before looking at the implications of these in the following chapters.

A **vector space** is defined as having the following mathematical objects:

1. An Abelian group $\langle V,+\rangle$ with an identity element e. We call members of the set V **vectors**. We define the identity element to be the zero vector, and we denote this by **0**. The operation + is called **vector addition**.
2. A field $\{F, +, \cdot\}$. We say that V is a vector space over the field F, and we call the members of F **scalars**.

> **The Zero Vector Is Not Denoted by $|0\rangle$**
>
> It is important to note that we denote the zero vector with a bold zero – **0** – and it is totally different from the vector $|0\rangle$ we defined earlier in the book. This is the convention in quantum computing.

We can define an additional operation as **scalar multiplication**, which is an operation between a scalar and a vector defined as follows:

- Let a scalar $s \in F$ and the vector $|v\rangle \in V$. Scalar multiplication is a binary operation, $f\colon F \times V \rightarrow V$. The multiplicative identity element of the field F of scalars is 1: $|v\rangle \cdot 1 = 1 \cdot |v\rangle = |v\rangle$.

This new operation – scalar multiplication – must also be compatible or work with addition and multiplication from our field F of scalars in rule two. It also has to be compatible with the operation of the vector addition defined in rule one. More formally, let α, $\beta \in F$ and $|u\rangle$, $|v\rangle \in V$. Or in other words, α and β are scalars and $|u\rangle$ and $|v\rangle$ are vectors:

- Scalar multiplication is compatible with field multiplication:
  $\alpha(\beta|v\rangle) = (\alpha\,\beta)|v\rangle$
- Distributivity of scalar multiplication with respect to vector addition:
  $\alpha(|u\rangle + |v\rangle) = \alpha|u\rangle + \alpha|v\rangle$
- Distributivity of scalar multiplication with respect to field addition:
  $(\alpha + \beta)|v\rangle = \alpha|v\rangle + \beta|v\rangle$

These two mathematical objects – an Abelian group $\langle V, +\rangle$ of vectors and a field $\{F, +, \cdot\}$ of scalars along with the operation of scalar multiplication – are all we need to define a vector space! That's it! That wasn't too bad, was it?

Any guesses which field our vector spaces in quantum computing are concerned with? If you answered the field of complex numbers, $\mathbb{C}$, give yourself three extra bonus points! However, for the next few chapters, I will stick to the field of real numbers, $\mathbb{R}$, as this makes it easier to get the concepts across without getting caught up in the extraneous complexities inherent in $\mathbb{C}$. Don't worry, there will be a whole chapter on complex numbers, and the latter half of the book will use this field almost exclusively.

Also, since vectors are the name we give to members of the set V, they can be *any* mathematical object. In quantum computing, they are n-tuples, but in mathematics and quantum mechanics, they can be anything, including functions, matrices, and polynomials. You do not need to worry about this for quantum computing, but you should know about it, as it will clarify why definitions are given in a generalized way to accommodate all of these possible things being vectors in a vector space.

Summary

In this chapter, we have built a solid foundation that will carry us through the rest of this book. We started with two fundamental mathematical concepts: sets and functions. From there, we defined a binary operation as being a function with two input values from sets. We combined all of these concepts to create groups, fields, and our ultimate goal, vector spaces. In the next chapter, we will look at all the great things we can do with these vectors that live in vector spaces!

Answers to Exercises

Exercise 1

Parts B and D are not functions.

Exercise 2

No, this is not a field. $\langle\mathbb{R}, -\rangle$ is not an Abelian group; subtraction is not commutative.

Works cited

- *"NumberSetinC"* by HB is licensed under CC BY-SA 4.0.
- *File:*`Cartesian Product qtl1.svg` - *Wikimedia Commons.*
- *"Cartesian Coordinate System"* by K. Bolino is licensed under CC BY-SA 3.0.
- *"Example Function"* by Bin im Garten is licensed under CC BY-SA 3.0.
- *"Example of Illegal Function"* by Bin im Garten is licensed under CC BY-SA 3.0.

4
Vector Spaces

The entire last chapter led up to defining vector spaces. Now we will see how some vector spaces can be subsumed by other vector spaces. We'll revisit linear combinations to talk about linear independence. We will also learn new ways to define vector spaces with just a small set of vectors. Finally, while we used the word dimension previously to describe a vector space, we will attain a mathematical definition for it in this chapter.

In this chapter, we are going to cover the following main topics:

- Subspaces
- Linear independence
- Span
- Basis
- Dimension

Subspaces

Let's say you have a set, U, of vectors and it is a subset of a set, V, of vectors ($U \subseteq V$). This situation is shown in the following diagram:

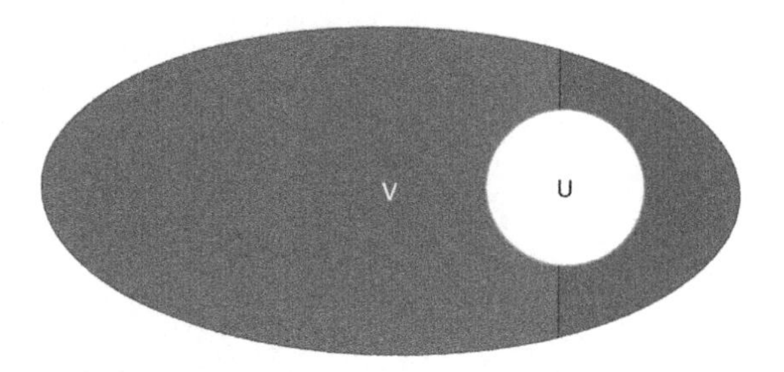

Figure 4.1 – The set U as a subset of V

Is it possible that U is a **subspace** of V? Well, yes. It has met the first condition for subspaces, namely, that the potential subspace has to be a subset of the bigger vector space's set of vectors. What's next? Well, U also has to be a vector space using the same field as the vector space V. This seems like it might be an exhaustive thing to do, but it has been proven that we only need to check for three small conditions to make sure U is a subspace of V, and two of them have to do with the closure property from *Chapter 3, Foundations*. As a reminder, here it is:

- **Closure**: For every $a, b \in A$, $a \circledast b$ produces an element, c, that is also in the set A. $f: A \times A \rightarrow A$

Armed with the concept of a subset and closure, we can now define a subspace.

Definition

For a subset, U, of a vector space, V, with an associated field, F, of scalars, U is a subspace if:

1. The **0** vector is included in the subset U. ($\mathbf{0} \in U$).
2. The subset U is closed for vector addition. If $|u\rangle$ and $|v\rangle \in U$, then $|u\rangle + |v\rangle \in U$.
3. The subset U is closed for scalar multiplication. If $|u\rangle \in U$ and s is a scalar, then $s|u\rangle \in U$.

We are basically trying to ensure that the following situation doesn't occur:

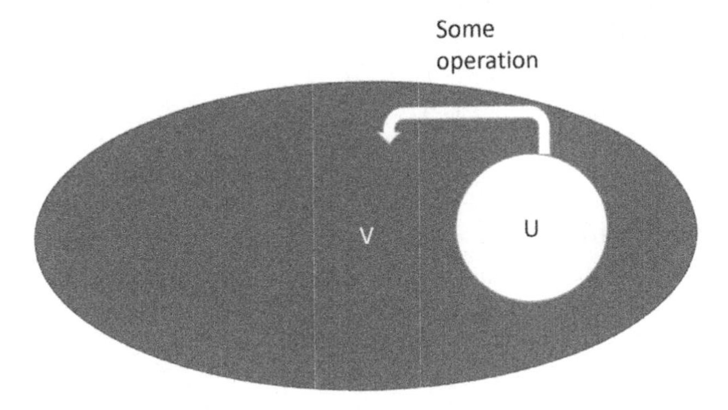

Figure 4.2 – Ensuring closure for a potential subspace

Alright, enough with formality! Let's look at concrete examples!

Examples

Let's use two-dimensional real space, $\mathbb{R}^2$, as our overall vector space to keep things easy. As you know, $\mathbb{R}^2$ is the complete Cartesian coordinate system. What if we picked a line on that X-Y plane to possibly be a subspace, shown as follows:

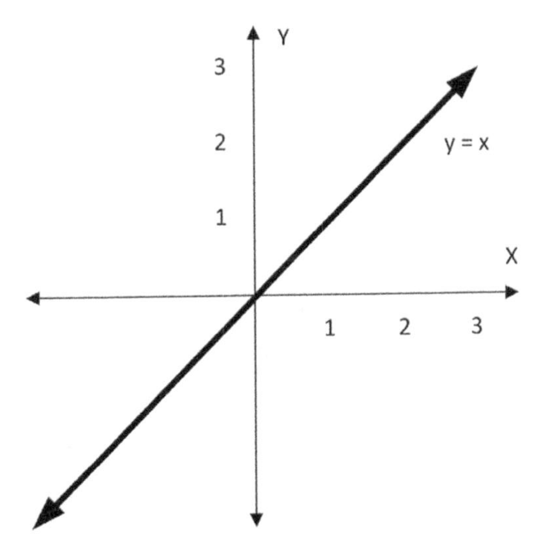

Figure 4.3 – The graph of $y = x$

As you can see, we picked the line $y = x$. Let's call every vector on that line (for example, (1,1), (2,2), (-3, -3)) the set, W, of vectors. Let's represent the set in our set builder notation from *Chapter 3, Foundations*:

$$W = \left\{ \begin{bmatrix} a \\ b \end{bmatrix} : a, b \in \mathbb{R}, a = b \right\}$$

Equivalently, we could write:

$$W = \left\{ \begin{bmatrix} a \\ a \end{bmatrix} : a \in \mathbb{R} \right\}$$

Using our graph, we can see that our set, W, contains a subset of $\mathbb{R}^2$ but is not equal to $\mathbb{R}^2$ since vectors such as (1,3) and (2,3) are not in our set. Now, let's check whether W is a *subspace* of $\mathbb{R}^2$.

1. The **0** vector is included in the subset U. ($\mathbf{0} \in U$).

   Check! The **0** vector in $\mathbb{R}^2$ is (0,0), and that meets the condition of our subset that the coordinates equal one another; so we're good here!

2. The subset U is closed for vector addition. If $|u\rangle$ and $|v\rangle \in U$, then $|u\rangle + |v\rangle \in U$.

   Hmmm, we'll have to work this one out:

$$\begin{bmatrix} x \\ x \end{bmatrix} + \begin{bmatrix} y \\ y \end{bmatrix} = \begin{bmatrix} x+y \\ x+y \end{bmatrix}$$

   It should be obvious that

$$x + y = x + y \ ,$$

   so we're all good and we can check this condition off the list.

3. The subset U is closed for scalar multiplication. If $|u\rangle \in U$ and s is a scalar, then $s|u\rangle \in$ U. Alright, let's work this one out too:

$$s\begin{bmatrix} a \\ a \end{bmatrix} = \begin{bmatrix} sa \\ sa \end{bmatrix}$$

   Again, it should be obvious that

$$sa = sa \ ,$$

   so, baa-ding! Check again!

So, all three conditions were true, and therefore W is a subspace of $\mathbb{R}^2$.

Let's try another line on the X-Y plane like the one in the following diagram:

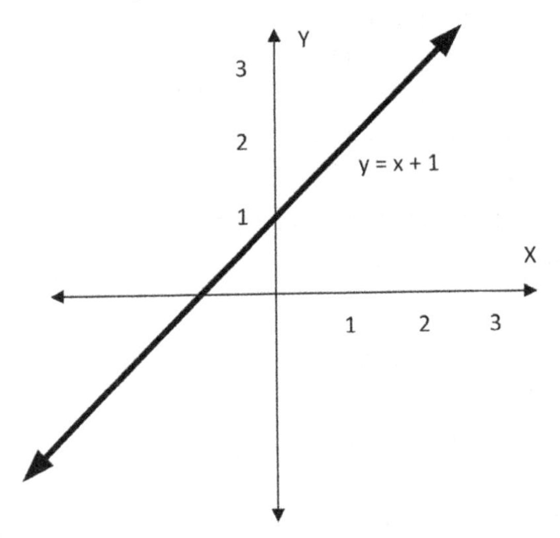

Figure 4.4 – The graph of $y = x + 1$

So, what do you think; is this one a subspace of $\mathbb{R}^2$? I can tell you right now, the answer is no. It fails the very first condition: it does not contain the zero vector. Alrighty, now it's your turn.

Exercise 1

Test the following subsets of $\mathbb{R}^2$ and determine which ones are subspaces and which are not. Assume all variables are real numbers.

1. $\left\{ \begin{bmatrix} a \\ b \end{bmatrix} : a > 0 \right\}$

2. $\left\{ \begin{bmatrix} 0 \\ a \end{bmatrix} \right\}$

3. $\left\{ \begin{bmatrix} a \\ a \end{bmatrix} : a \geq 0 \right\}$

Linear independence

So, it ends up that these vectors got together and wrote a declaration of independence and that's what we'll cover here. Just joking! We do need humor every so often in a math book. To explain **linear independence,** we need to go back to the concept of a linear combination that we introduced earlier in this book.

Linear combination

We learned in *Chapter 2, Superposition with Euclid*, that linear combinations are the scaling and addition of vectors. I would like to give a more precise definition as we go beyond three-dimensional space.

A linear combination for vectors $|x_1\rangle, |x_2\rangle, \ldots |x_n\rangle$ and scalars $c_1, c_2, \ldots c_n$ in a vector space, V, is a vector of the form:

$$c_1 |x_1\rangle + c_2 |x_2\rangle + \cdots + c_n |x_n\rangle \qquad (1)$$

Basically, it is still scaling and addition, but now we can do it for vectors of any dimension and with as many finite numbers of vectors as we wish.

Let's look at an example:

$$2\begin{bmatrix} 1 \\ 0 \\ -1 \\ 2 \end{bmatrix} + 3\begin{bmatrix} -1 \\ 2 \\ -2 \\ 0 \end{bmatrix} - 2\begin{bmatrix} 2 \\ 3 \\ -2 \\ 1 \end{bmatrix} = \begin{bmatrix} 2 \\ 0 \\ -2 \\ 4 \end{bmatrix} + \begin{bmatrix} -3 \\ 6 \\ -6 \\ 0 \end{bmatrix} + \begin{bmatrix} -4 \\ -6 \\ 4 \\ -2 \end{bmatrix} = \begin{bmatrix} -5 \\ 0 \\ -4 \\ 2 \end{bmatrix}$$

So now that we have defined linear combinations, let's look at the antonym of linear independence – **linear dependence**.

Linear dependence

If you have a set of vectors and you can create a linear combination of one of the vectors from a subset of the other vectors, then all of those vectors are linearly dependent. Let's look at some examples.

The following vectors are linearly dependent:

$$|d\rangle = \begin{bmatrix} 1 \\ 2 \\ 3 \end{bmatrix} \quad |e\rangle = \begin{bmatrix} 5 \\ 6 \\ 7 \end{bmatrix} \quad |f\rangle = \begin{bmatrix} 13 \\ 18 \\ 23 \end{bmatrix}$$

This is because we can create $|f\rangle$ from a linear combination of $|d\rangle$ and $|e\rangle$ using the scalars *3* and *2*, as shown here:

$$3|d\rangle + 2|e\rangle = \begin{bmatrix} 3 \\ 6 \\ 9 \end{bmatrix} + \begin{bmatrix} 10 \\ 12 \\ 14 \end{bmatrix} = \begin{bmatrix} 13 \\ 18 \\ 23 \end{bmatrix}$$

Are $|0\rangle$ and $|1\rangle$ linearly dependent? No, because there are no scalars that you can multiply $|0\rangle$ by to get $|1\rangle$ and vice-versa. You should quickly verify this.

How about these vectors?

$$|g\rangle = \begin{bmatrix} 5 \\ -7 \\ 3 \\ 1 \end{bmatrix} \quad |h\rangle = \begin{bmatrix} -10 \\ -3 \\ 6 \\ 8 \end{bmatrix} \quad |f\rangle = \begin{bmatrix} 4 \\ 8 \\ 9 \\ -3 \end{bmatrix}$$

These are not linearly dependent as there are no scalars that you can use to create one out of a subset of all three. Now, you might rightly ask, how do you know? That is a very good question and there are whole branches of computational linear algebra dedicated to just this problem. The short answer is that there are several methods, some easier than others. However, I'm going to give you a pass. The methods are not needed for quantum computing, just the concept. You can use a very nice calculator to do the computation for you, such as **Wolfram Alpha** (`https://www.wolframalpha.com/`). Here is a screenshot from there asking about those three vectors we just defined:

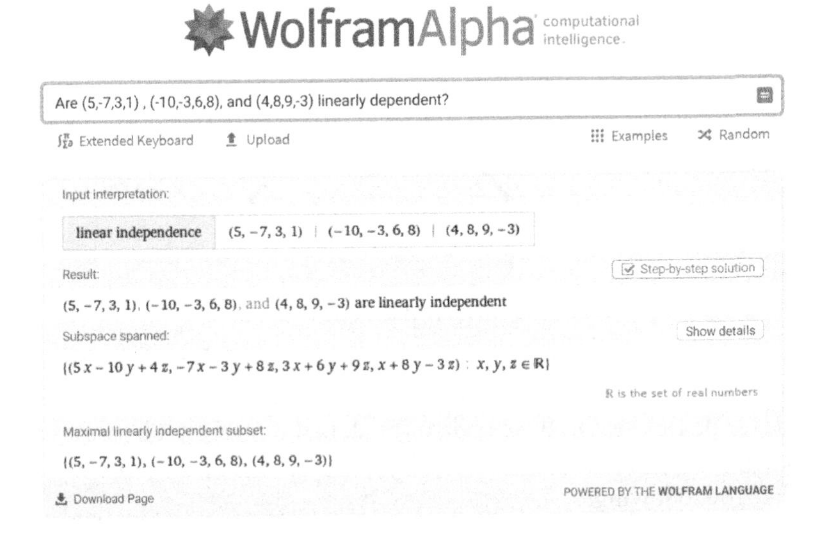

Figure 4.5 – Screenshot from Wolfram Alpha

If you read the screenshot carefully, it gives you a hint to this next part. A set of vectors is either linearly dependent or linearly independent! They cannot be both and they have to be one or the other. In fact, that is how we will define linear independence:

A set of vectors is **linearly independent** *if they are not linearly dependent.*

So, there you have it, you now know what linear independence is. Be sure not to forget this concept as it will be important later on.

Span

In the section on subspaces, we used set builder notation to define possible candidates for subspaces. There is a better way to do this, however, using something called the span. The **span** uses a set of distinct, indexed vectors to generate a vector space. How does it do this? It uses every possible linear combination of the set of vectors. As you hopefully see by now, linear combinations are at the heart of linear algebra.

So, let's start with a set, *S*, of vectors with just one vector, like the following:

$$S = \left\{ \begin{bmatrix} 2 \\ 2 \end{bmatrix} \right\}$$

Let's look at our one vector graphically:

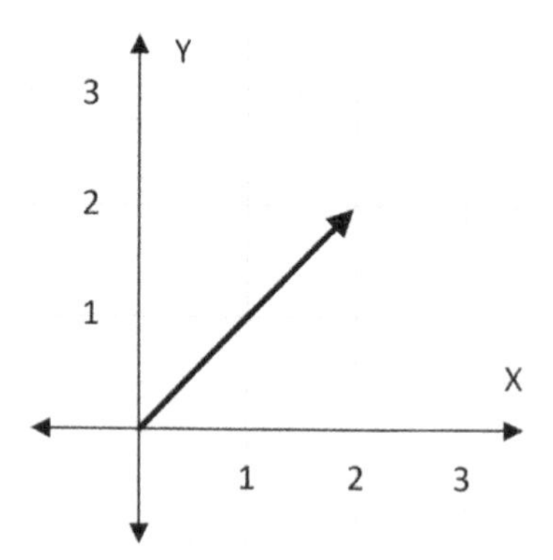

Figure 4.6 – Graph of the one vector in S

Okay, what would be its span? Or, in other words, what are all the vectors that are linear combinations of this one vector? Well, we can't add because all we have is one vector. So all we can do is scale this vector. If we scale it for all real numbers, it becomes a ..... line! We would say that *S* spans the space, or span(*S*) generates the space. So, our subspace would look like the following:

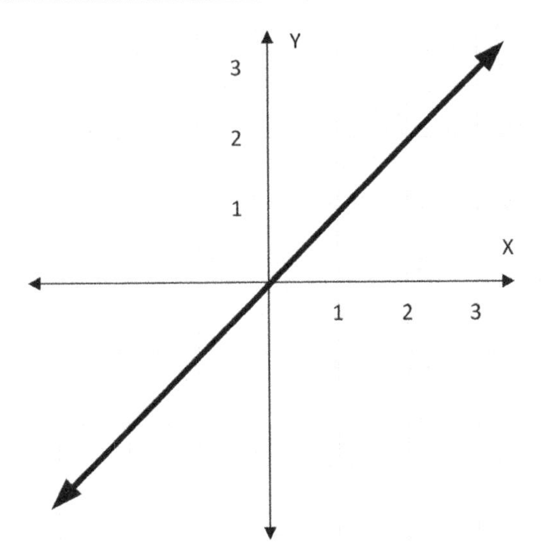

Figure 4.7 – Graph of the span of S

Hopefully, this line looks familiar, it is the line $y = x$ from the previous section. So, we have just created a subspace of $\mathbb{R}^2$ with one vector using the span operator.

Let's add a vector and see whether we can generate a more interesting subspace:

$$T = \left\{ \begin{bmatrix} 2 \\ 2 \end{bmatrix}, \begin{bmatrix} -2 \\ -2 \end{bmatrix} \right\}$$

Let's look at these vectors graphically:

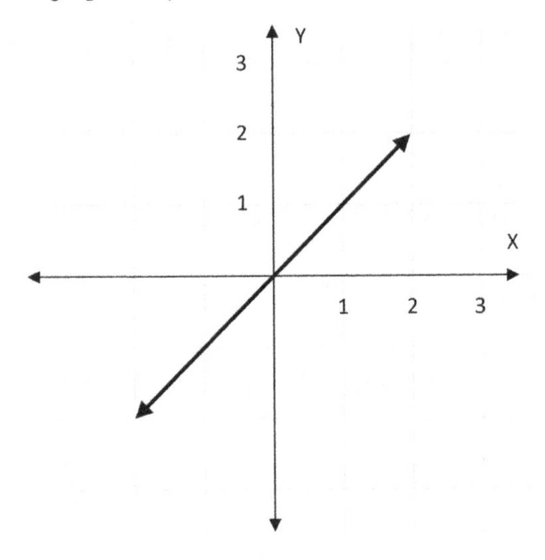

Figure 4.8 – Graph of the two vectors in set T

As you can see, they lie on the same line. Now, the big question is, what are all their linear combinations? The answer ends up being disappointing. Yes, we do have another vector to add now, but since the two vectors are *linearly dependent*, it buys us nothing. All their linear combinations end up being the same line we had before, $y = x$. So, $\text{span}(S) = \text{span}(T)$.

To really get something, we need to add vectors that are *linearly independent* to the vectors already in our spanning set. Let's go ahead and do that:

$$U = \left\{ \begin{bmatrix} 2 \\ 2 \end{bmatrix}, \begin{bmatrix} -2 \\ -2 \end{bmatrix}, \begin{bmatrix} 0 \\ 1 \end{bmatrix} \right\}$$

The third vector is not a linear combination of the first two. Let's draw them on a graph:

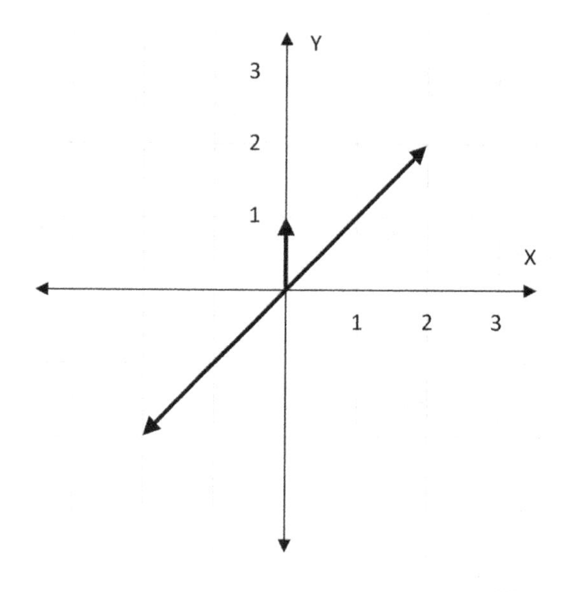

Figure 4.9 – Graph of the three vectors in set U

Alright, now we can do something! What are all the linear combinations of the three vectors in our set *U*? I've drawn a few as follows:

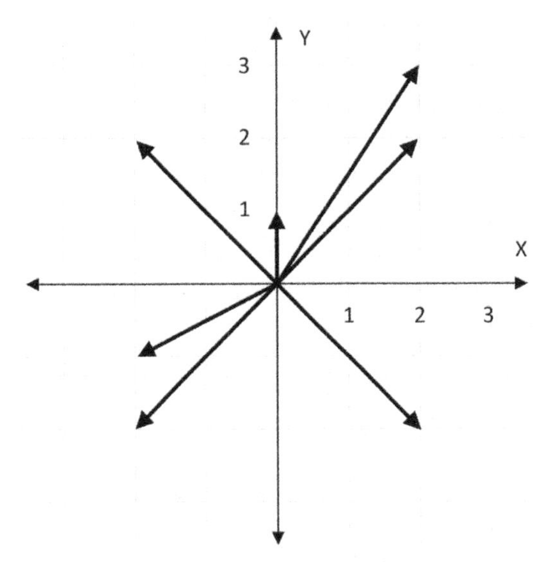

Figure 4.10 – Graph of the linear combinations of vectors in the set U

If we continued to draw all the vectors that are linear combinations of the set *U*, we would end up filling the entire vector space of $\mathbb{R}^2$! That's right; we just defined $\mathbb{R}^2$ with only three vectors:

$$span(U) = \mathbb{R}^2$$

Can we find a set with fewer vectors that spans $\mathbb{R}^2$? It just so happens that we can! The following set will do the job:

$$W = \left\{ \begin{bmatrix} 0 \\ 3 \end{bmatrix}, \begin{bmatrix} 3 \\ 3 \end{bmatrix} \right\}$$

This set of vectors is *linearly independent*. This condition of linear independence for a spanning set ends up being so important that we give spanning sets of linearly independent vectors a special name – a **basis**.

Basis

The word basis is used often in English speech and its colloquial definition is actually a good way to look at the word basis in linear algebra:

> **Basis, ba·sis \ 'bā-səs \ *plural* bases\ 'bā-ˌsēz \ Noun**
> Something on which something else is established or based. Example 1: Stories with little basis in reality. Example 2: No legal basis for a new trial.

The reason for this is that you can choose different bases for a vector space. While the vector space itself does not change when you choose a different basis, the way things are described with numbers does.

Let's look at an example in $\mathbb{R}^2$. Consider the vector $|u\rangle$, given as follows:

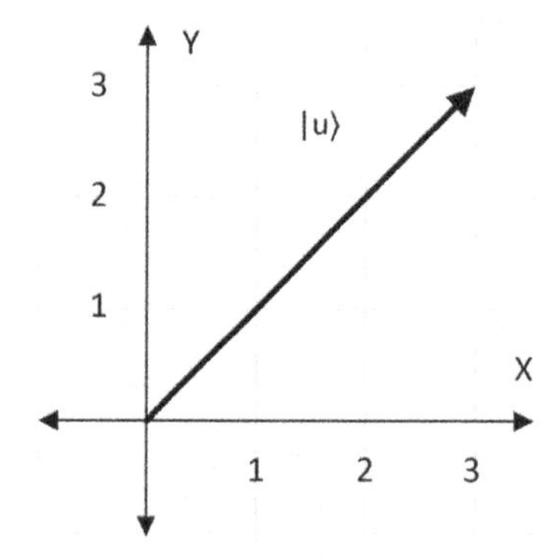

Figure 4.11 – Graph of the vector |u⟩

Clearly, its coordinates are (3,3). What if I told you I could describe the same vector with the coordinates (3,0)? Wait a minute; that should disturb you. We never talk about the basis in most math classes because we implicitly use the **standard basis**. What is the standard basis, you may ask? It is the set of vectors whose components are all zeros, except one component that has the number one. For $\mathbb{R}^2$, they are:

$$|0\rangle = \begin{bmatrix} 1 \\ 0 \end{bmatrix} \qquad |1\rangle = \begin{bmatrix} 0 \\ 1 \end{bmatrix}$$

You will notice that these vectors are our familiar zero- and one-state vectors. This is no accident. In quantum computing, we call the standard basis the **computational basis**, and that is the term we will use in this book.

Okay, now remember when I said you could write any vector as a linear combination of a spanning set? Well, since a basis is a spanning set, this is true for a basis as well. But what's special about a basis is that there is a *unique* linear combination to describe each vector and the scalars used in this unique linear combination are called **coordinates**.

Let's look at this for our example vector $|u\rangle$. The unique linear combination to describe $|u\rangle$ using the standard basis is:

$$x|0\rangle + y|1\rangle = \begin{bmatrix} 3 \\ 3 \end{bmatrix}$$

$$3\begin{bmatrix} 1 \\ 0 \end{bmatrix} + 3\begin{bmatrix} 0 \\ 1 \end{bmatrix} = \begin{bmatrix} 3 \\ 3 \end{bmatrix}$$

I purposely named the scalars x and y because they correspond to the x and y coordinates according to the computational basis.

Now, I am going to describe $\mathbb{R}^2$ with a different basis, F, shown as follows:

$$F = \{ |f_1\rangle, |f_2\rangle \} \; where \; |f_1\rangle = \begin{bmatrix} 1 \\ 1 \end{bmatrix} and \; |f_2\rangle = \begin{bmatrix} -1 \\ 1 \end{bmatrix}$$

Let's look at $|f_1\rangle$ and $|f_2\rangle$ on our standard coordinate system:

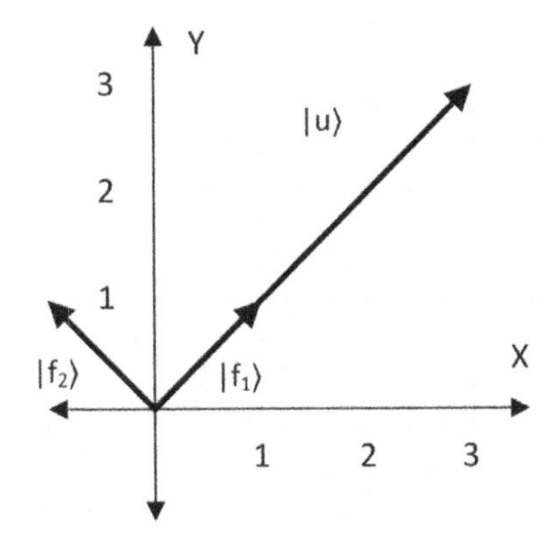

Figure 4.12 – Graph of the vectors in set F

This basis is just as valid as the computational basis from before because every vector in $\mathbb{R}^2$ can be uniquely defined by a linear combination of these two vectors. So, the next question is: What is the linear combination that describes $|u\rangle$? You can probably make out geometrically what the unique linear combination is that describes $|u\rangle$, but let's write it out algebraically as well:

$$x|f_1\rangle + y|f_2\rangle = \begin{bmatrix} 3 \\ 3 \end{bmatrix}$$

$$3\begin{bmatrix} 1 \\ 1 \end{bmatrix} + 0\begin{bmatrix} -1 \\ 1 \end{bmatrix} = \begin{bmatrix} 3 \\ 3 \end{bmatrix}$$

So, the scalars for this linear combination are (3,0) and these are the coordinates of $|u\rangle$ according to our basis, F. Let's look at $|u\rangle$ in our new coordinate system:

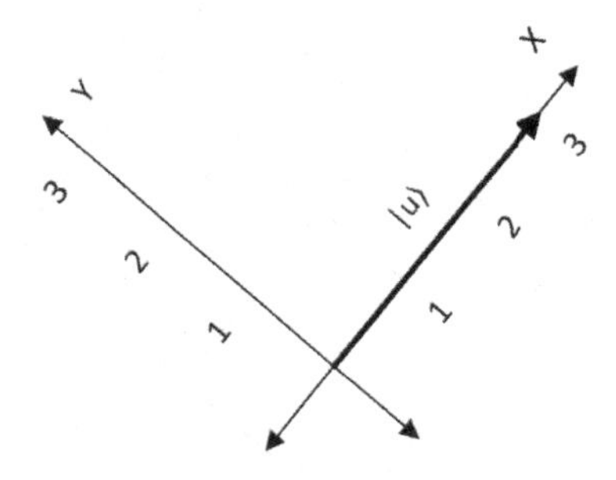

Figure 4.13 – Graph of $|u\rangle$ in the F basis

Notice that $|u\rangle$ is still the same vector according to our very first definition of a vector as being a line segment with a length and direction. But with a new basis, we just use different coordinates to describe it. As Shakespeare wrote in Romeo and Juliet:

What's in a name? That which we call a rose

By any other name would smell as sweet;

Or, in other words:

What's in coordinates? That which we call a vector

By any other coordinates would still be the same vector;

So, I have now shown you how I can describe the same vector with different components or coordinates. To denote the basis we are writing a vector in, we use a subscript like so, where C denotes the computational basis:

$$\begin{bmatrix} 3 \\ 3 \end{bmatrix}_C = \begin{bmatrix} 3 \\ 0 \end{bmatrix}_F$$

If no basis is given, then we assume we are using the computational basis. It should be noted that the basis has to be indexed and ordered so that our coordinates do not get mixed up (for example, (0,3) does not equal (3,0) in the computational basis).

I know all of this may be a little confusing, but I need to blow your mind a little to prepare you for what lies ahead. In the next chapter, I will show you how to use matrices to travel between bases easily.

Dimension

The **dimension** of a vector space ends up being very easy to define once you know what a basis is. One thing we didn't talk about in our previous section is that the basis is the *minimum* set of vectors needed to span a space, and that the number of vectors in a basis for a particular vector space is always the same.

From this, we define the dimension of a vector space to be equal to the number of vectors it takes as a basis to describe a vector space. Equivalently, we could say that it is the number of coordinates it takes to describe a vector in the vector space. It follows that the dimension of $\mathbb{R}^2$ is two, $\mathbb{R}^3$ is three, and $\mathbb{R}^n$ is n.

Summary

We have covered a bit of ground around describing vector spaces in this chapter. We've seen how a vector space has subspaces and how to test whether a set of vectors is a subspace. We've rigorously defined linear combinations and derived the concept of linear independence from it. We've also learned multiple ways to describe a vector space through the span and basis. From this, we've learned the true meaning of coordinates and put all that together to define the dimension of a vector space. In the next chapter, we will look at how to transform vectors in these vector spaces using matrices!

Answers to exercises

Exercise 1

1. No, it is not a subspace. It does not contain the zero vector.
2. Yes, it is a subspace!
3. No, it is not a subspace. It is not closed under scalar multiplication.

5 Using Matrices to Transform Space

Linear transformations are one of the most central topics in linear algebra. Now that we have defined vectors and vector spaces, we need to be able to do things with them. In *Chapter 3, Foundations*, we manipulated mathematical objects with functions. When we manipulate vectors in vector spaces, mathematicians use the term *linear transformations*. Why the change of terminology? As with most things in linear algebra, the wording is inspired by Euclidean geometry. We will see that, geometrically, these "functions" actually "transform" vectors from one direction and length to another. But this visual transformation has been generalized algebraically to all types of vectors (n-tuples of numbers, functions, and so on).

We also go through the crucial link between linear transformations and matrices. The most important point of this chapter is that linear transformations can *always* be represented by matrices when the vector spaces are finite (which are the only ones we use in this book). The only caveat is that this is not a one-to-one relationship but rather a one-to-many relationship in that each linear transformation can be represented by multiple matrices.

A word of caution about terminology. Some mathematicians call a linear transformation a **linear mapping**. Physicists and quantum computing practitioners use the term **linear operator**, which we will consider as a special type of linear transformation at the end of this chapter.

In this chapter, we are going to cover the following main topics:

- Linearity
- What is a linear transformation?
- Representing linear transformations with matrices
- Transformations inspired by Euclid
- Linear operators
- Linear functionals
- A change of basis

Linearity

What makes a transform linear? This question gets to the heart of linear algebra. The concept of linearity ties together all the other concepts we have considered so far and the ones to come. Indeed, quantum mechanics is a *linear* theory. That's what makes linear algebra crucial to understanding quantum computing.

Before I define linearity, let's look at what it is not. Real-life examples of non-linearity abound. For example, exercising 1 hour a day for 24 days does not give the same result as exercising 24 hours in 1 day. Watering a plant is another good non-linear example. Giving a plant 1 gallon of water a day for 100 days will be much better than giving it 100 gallons in 1 day. These are both examples of non-linear relationships. How much you put in does not always translate to what you get out.

Linear relationships, on the other hand, are proportional. Speed is a good example. If you go 20 mph for 1 hour, you'll cover 20 miles. If you go 1 mph for 20 hours, you'll still cover 20 miles. Exchange rates for money are also a good example. If the current rate is 2 dollars to a euro, then if I give you 4 dollars, you'll give me 2 euros. I can also give you 1 dollar 4 times and get the same result.

Graphs for these types of relationships are straight lines, such as this one for the euro example:

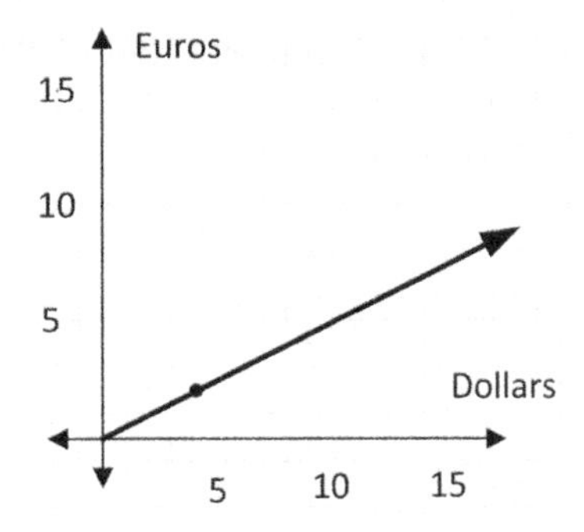

Figure 5.1 – A graph of the dollars to euros function

Crucially though, linearity requires functions to return 0 if 0 is the input to the function. So, straight-line functions that don't pass through the origin do not get ascribed the property of linearity. In other words, if I go 0 mph, I should cover 0 miles, and hopefully, you won't give me euros back if I give you 0 dollars (but you'd be my best friend if you did).

In mathematics, we want to generalize this concept so that it works in many situations. We are all familiar with a line through the origin. Let's take the simplest example, $y = x$, as shown in the following graph:

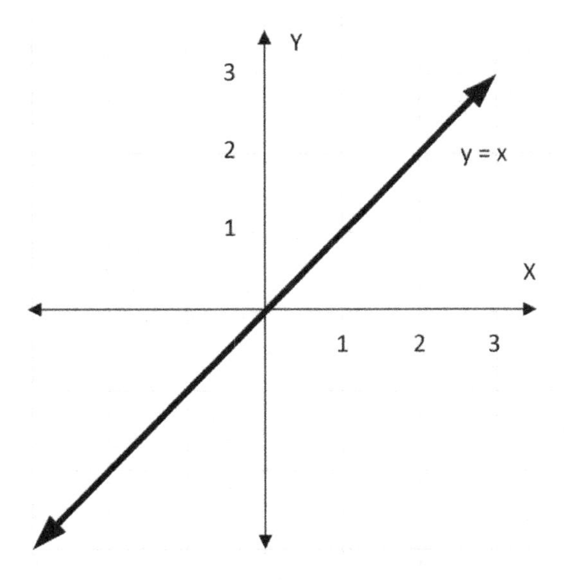

Figure 5.2 – The line $y = x$

What can we generalize about this line? Well, the line has a constant slope, namely one, so that if I increase its slope, it consequently raises the output by a proportional amount. For example, let's say I increase the slope to three ($y = 3x$). Well, now instead of y being 3 at $x = 1$, it will be $3 \cdot 3$ or 9. This property has been generalized into something called **homogeneity** and is defined thusly:

$$\alpha f(x) = f(\alpha x) \text{ where } \alpha \text{ is a scalar}$$

Here is a small table showing values for our function $y = x$ and the different values of α:

x	α	f(x)	f(α x)	α f(x)
-2	-1	-2	2	2
-1	1	-1	-1	-1
0	5	0	0	0
1	2	1	2	2

Table 5.1 – αf(x) and f(αx) at different values of x and α

What else can we generalize about our line $y = x$? Well, if I added the value of y at $x = 3$ to the value of y at $x = 4$, it would equal the value of y at $x = 7$. This works no matter what slope I give the line, too. Here's another table of values to prove it to you:

x	*z*	*f(x)*	*f(z)*	*f(x + z)*
-2	-1	-2	-1	-3
-1	1	-1	1	0
0	5	0	5	5
1	2	1	2	3

Table 5.2 – f(x), f(z), and f(x+z) at different values of x and z

This property is called **additivity** and is defined this way:

$$f(x+y)=f(x)+f(y)$$

These two properties, additivity and homogeneity, define **linearity**. To be a *linear* transformation, a transformation must have *linearity*.

What is a linear transformation?

To be precise, a linear transformation is a function T from a vector space U to a vector space V. A capital letter "T" is traditionally used by mathematicians to denote a generic transformation, and we use the same syntax that we introduced for functions in *Chapter 3, Foundations*:

$$T:U \to V$$

Similarly, the vector space U is the **domain**, and the vector space V is the **codomain**. Each vector that is transformed is called the **image** of the original vector in the domain. The set of all images is the **range**.

To be linear, the transformation must preserve the operations of vector addition and scalar multiplication by meeting the conditions for linearity. Here, we express them in terms of vectors:

$$T\left(|x\rangle + |y\rangle \right) = T\left(|x\rangle \right) + T\left(|y\rangle \right)$$
$$T\left(s|x\rangle \right) = sT\left(|x\rangle \right)$$
$$\text{where } s \text{ is a scalar}$$

It follows from these axioms that for any linear transformation T, $T(0)$ has to equal the zero vector 0. Let's look at how we describe transformations in the next section.

Describing linear transformations

There are many ways to describe a linear transformation. In the case of Euclidean vectors, you can describe a linear transformation geometrically. If you are dealing with n-tuples of numbers, you can describe the effect of the transformation on those numbers. Due to the concept of linear combinations, you can also describe how the transformation changes just the basis vectors. Let's go through each of these in depth.

A geometric description

To show an example of a geometric description, let's use reflection. Reflection is a very easy and intuitive transformation, as seen in the following diagram. We will call our reflection transformation "R":

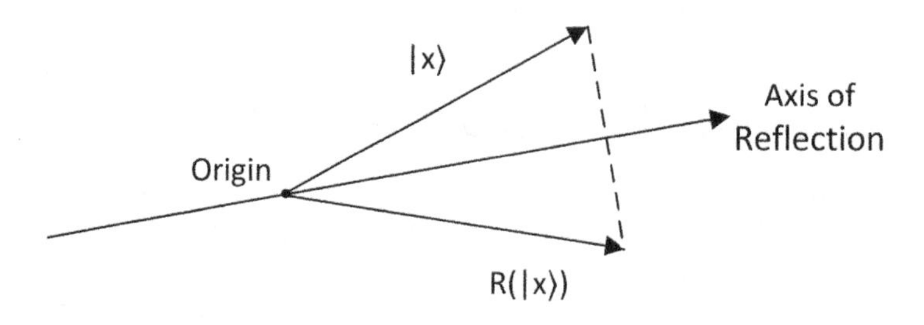

Figure 5.3 – A graphical depiction of the reflection transformation

One vector is the axis of reflection, and you reflect vectors about it by drawing a dashed line that is perpendicular to the axis of reflection from the tip of the original vector ($|x\rangle$ in the diagram). Then, place the tip of the reflected vector $R(|x\rangle)$ equidistant from the axis of reflection.

Okay, now that we've described this transformation, let's check to see whether it is linear. First, we'll check for homogeneity. Homogeneity for vector transformations is defined as:

$$T(s|x\rangle) = sT(|x\rangle)$$

We will set $s = 2$, and you can see in the following diagram that our reflection transformation, R, does indeed pass the test for homogeneity:

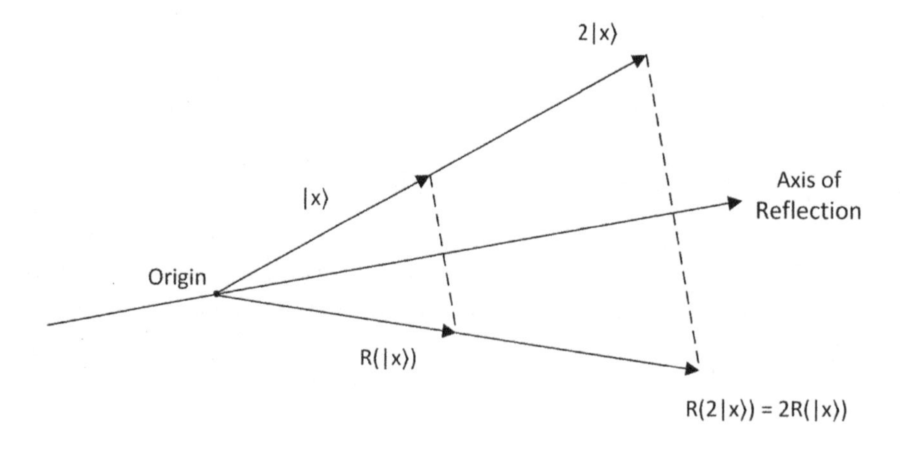

Figure 5.4 – A test reflection for homogeneity

Now, let's test the other condition for linearity, additivity. As you should recall, additivity for a linear transformation is defined as:

$$T(|x\rangle + |y\rangle) = T(|x\rangle) + T(|y\rangle)$$

The following diagram shows a reflection for two vectors, $|x\rangle$ and $|y\rangle$. Note that $R(|y\rangle)$ is the same as $|y\rangle$ because $|y\rangle$ is on the axis of reflection:

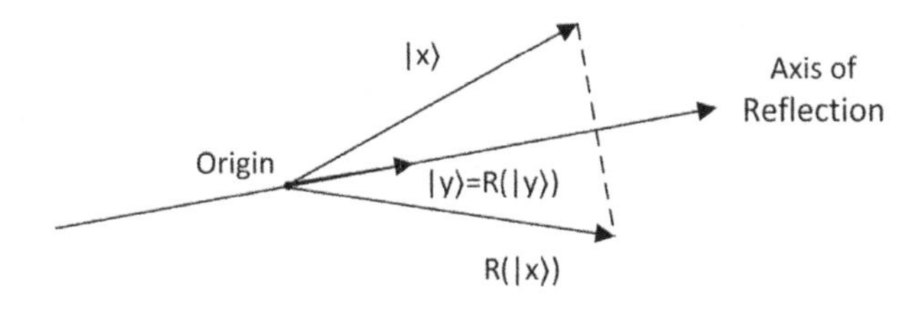

Figure 5.5 – A reflection of $|y\rangle$

Now, look at the following diagram closely:

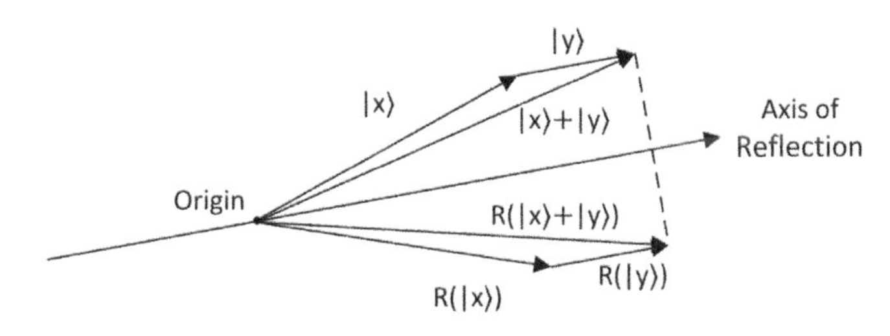

Figure 5.6 – A test for additivity

We moved the start point of $|y\rangle$and $R(|y\rangle)$ to the end of $|x\rangle$ and $R(|x\rangle)$, respectively, because Euclidean vectors are equal as long as they retain their length and magnitude, as we explained in *Chapter 1, Superposition with Euclid*. From the diagram, you should be able to make out that additivity holds for reflections as well. Therefore, our reflection transformation is a *linear* transformation.

An algebraic description

Let's look at another way you can describe a linear transformation. If you are dealing with n-tuples of numbers in $\mathbb{R}^n$, then you can say explicitly what the transform does to an n-tuple. Let's look at an example.

First, I need to define the domain and codomain of the transform, like so:

$$T : \mathbb{R}^2 \rightarrow \mathbb{R}^3$$

Then, I can define what the transform does:

$$T(x, y) = (2x + y, x - y, x + y)$$

We represent n-tuples as column vectors, so this can be rewritten as:

$$T(|x\rangle) = T\left(\begin{bmatrix} x \\ y \end{bmatrix}\right) = \begin{bmatrix} 2x + y \\ x - y \\ x + y \end{bmatrix}$$

The preceding equations fully define the linear transformation for every vector in our domain, $\mathbb{R}^2$. Here are a few instances of the transformation in action:

$$T\left(\begin{bmatrix} 2 \\ 3 \end{bmatrix}\right) = \begin{bmatrix} 7 \\ -1 \\ 5 \end{bmatrix}$$

$$T\left(\begin{bmatrix} -2 \\ 1 \end{bmatrix}\right) = \begin{bmatrix} -3 \\ -3 \\ -1 \end{bmatrix}$$

$$T\left(\begin{bmatrix} 6 \\ 3 \end{bmatrix}\right) = \begin{bmatrix} 15 \\ 3 \\ 9 \end{bmatrix}$$

We should make sure this transformation is indeed linear as well. Let's do homogeneity first:

$$T(s|x\rangle) = sT(|x\rangle)$$

Let's see how our transformation does with this:

$$T(s|x\rangle) = T\left(\begin{bmatrix} sx \\ sy \end{bmatrix}\right) = \begin{bmatrix} 2sx + sy \\ sx - sy \\ sx + sy \end{bmatrix}$$

$$sT(|x\rangle) = sT\left(\begin{bmatrix} x \\ y \end{bmatrix}\right) = s\begin{bmatrix} 2x + y \\ x - y \\ x + y \end{bmatrix} = \begin{bmatrix} 2sx + sy \\ sx - sy \\ sx + sy \end{bmatrix}$$

Alright, it passes! Let's try additivity:

$$T(|x\rangle + |y\rangle) = T(|x\rangle) + T(|y\rangle)$$

Here's the test:

$$T(|x\rangle+|y\rangle) = T\left(\begin{bmatrix} x \\ y \end{bmatrix} + \begin{bmatrix} w \\ z \end{bmatrix}\right) = T\left(\begin{bmatrix} x+w \\ y+z \end{bmatrix}\right)\begin{bmatrix} 2(x+w)+(y+z) \\ (x+w)-(y+z) \\ (x+w)+(y+z) \end{bmatrix} = \begin{bmatrix} 2x+2w+y+z \\ x+w-y-z \\ x+w+y+z \end{bmatrix}$$

$$T(|x\rangle) + \mathrm{T}(|y\rangle) = T\left(\begin{bmatrix} x \\ y \end{bmatrix}\right) + T\left(\begin{bmatrix} w \\ z \end{bmatrix}\right) = \begin{bmatrix} 2x+y \\ x-y \\ x+y \end{bmatrix} + \begin{bmatrix} 2w+z \\ w-z \\ w+z \end{bmatrix} = \begin{bmatrix} 2x+y+2w+z \\ x-y+w-z \\ x+y+w+z \end{bmatrix}$$

Again, it passes, so this transformation is linear. Now it's your turn – are the following transforms linear?

Exercise one

$$T:\mathbb{R}^4 \to \mathbb{R}^3$$

$$T(|x\rangle) = T\left(\begin{bmatrix} x \\ y \\ w \\ z \end{bmatrix}\right) = \begin{bmatrix} w \\ y \\ z \end{bmatrix}$$

$$U:\mathbb{R}^2 \to \mathbb{R}^3$$

$$U(|x\rangle) = U\left(\begin{bmatrix} x \\ y \end{bmatrix}\right) = \begin{bmatrix} x+y \\ y \\ x^2 \end{bmatrix}$$

$$Y:\mathbb{R}^3 \to \mathbb{R}^3$$

$$Y(|x\rangle) = Y\left(\begin{bmatrix} x \\ y \\ z \end{bmatrix}\right) = \begin{bmatrix} x+y \\ 2y \\ z+2 \end{bmatrix}$$

A basis vectors description

I'd like to show you one more way to describe a linear transformation. There are many more ways to describe a linear transformation, but I think these three are a good way to start.

Since any vector in a vector space can be expressed as a linear combination of a set of basis vectors, if you describe what the transform does to a set of basis vectors, you've described the complete transformation. Let's show this through an example.

We'll start with our computational basis vectors $|0\rangle$ and $|1\rangle$. I'll describe what my transform does to these two vectors:

$$T:\mathbb{R}^2 \rightarrow \mathbb{R}^2$$

$$T(|0\rangle) = \begin{bmatrix} 1 \\ 2 \end{bmatrix} \quad , \quad T(|1\rangle) = \begin{bmatrix} 2 \\ 1 \end{bmatrix}$$

I have now fully described the transformation for every vector in my domain $\mathbb{R}^2$. Let's take a random vector, $|x\rangle$ in $\mathbb{R}^2$, and work out its transformation:

$$|x\rangle = \begin{bmatrix} 3 \\ 2 \end{bmatrix} = 3|0\rangle + 2|1\rangle$$

You'll notice that I've expressed the vector $|x\rangle$ as a linear combination of our basis vectors. By definition of a basis, I can do this for any vector in $\mathbb{R}^2$. Now, I will apply the transformation to the linear combination:

$$T(|x\rangle) = T\left(\ 3|0\rangle + 2|1\rangle\ \right)$$

Due to Additivity, I can do this:

$$T\left(\ 3|0\rangle + 2|1\rangle\ \right) = T(3|0\rangle) + T(2|1\rangle)$$

Due to Homogenity, I can do this:

$$T(3|0\rangle) + T(2|1\rangle) = 3T(|0\rangle) + 2T(|1\rangle)$$

I can then substitue the values of $T(|0\rangle)$ and $T(|1\rangle)$ in:

$$3T(|0\rangle) + 2T(|1\rangle) = 3\begin{bmatrix} 1 \\ 2 \end{bmatrix} + 2\begin{bmatrix} 2 \\ 1 \end{bmatrix} = \begin{bmatrix} 3 \\ 6 \end{bmatrix} + \begin{bmatrix} 4 \\ 2 \end{bmatrix} = \begin{bmatrix} 7 \\ 8 \end{bmatrix}$$

$$\text{So } T(|x\rangle) = \begin{bmatrix} 7 \\ 8 \end{bmatrix}.$$

Through this example, I hope I've shown that you can describe a linear transformation by just stating what it does to a set of basis vectors. We'll move on now to matrices!

Representing linear transformations with matrices

Now for the most common and important way of describing a linear transformation, the matrix. Through the magic of matrix-vector multiplication, a matrix is all you need to describe a linear transformation.

Again, let's start with an example. I'm going to describe the linear transformation we used in the *An algebraic description* section with a matrix. To jog your memory, here is the aforementioned linear transformation:

$$T:\mathbb{R}^2 \to \mathbb{R}^3$$

$$T(|x\rangle) = T\left(\begin{bmatrix} x \\ y \end{bmatrix}\right) = \begin{bmatrix} 2x+y \\ x-y \\ x+y \end{bmatrix}$$

Now, here is how I can describe it with a matrix:

$$T:\mathbb{R}^2 \to \mathbb{R}^3$$

$$T(|x\rangle) = T\left(\begin{bmatrix} x \\ y \end{bmatrix}\right) = \begin{bmatrix} 2 & 1 \\ 1 & -1 \\ 1 & 1 \end{bmatrix}\begin{bmatrix} x \\ y \end{bmatrix}$$

I don't even need to be that formal, other than telling you that we are using real numbers; I can just give you the matrix, and that describes everything. The dimension of the domain is the number of columns of the matrix, the dimension of the codomain is the number of rows of the matrix, and the actual transformation is the matrix itself. That is the power of a matrix!

Let's apply this transformation to the same example vectors we used in the *An algebraic description* section using matrix-vector multiplication:

$$T\left(\begin{bmatrix}2\\3\end{bmatrix}\right)=\begin{bmatrix}2 & 1\\1 & -1\\1 & 1\end{bmatrix}\begin{bmatrix}2\\3\end{bmatrix}=\begin{bmatrix}7\\-1\\5\end{bmatrix}$$

$$T\left(T\begin{bmatrix}-2\\1\end{bmatrix}\right)=\begin{bmatrix}2 & 1\\1 & -1\\1 & 1\end{bmatrix}\begin{bmatrix}-2\\1\end{bmatrix}=\begin{bmatrix}-3\\-3\\-1\end{bmatrix}$$

$$T\left(\begin{bmatrix}6\\3\end{bmatrix}\right)=\begin{bmatrix}2 & 1\\1 & -1\\1 & 1\end{bmatrix}\begin{bmatrix}6\\3\end{bmatrix}=\begin{bmatrix}15\\3\\9\end{bmatrix}$$

We get the same exact answers!

Matrices depend on the bases chosen

Now, let's look to see how we can use a matrix to describe our reflection transformation from before. The crucial point of this example is that our matrix depends on the basis we choose to represent it. Look at the transformation and see whether you can determine any good basis to use it:

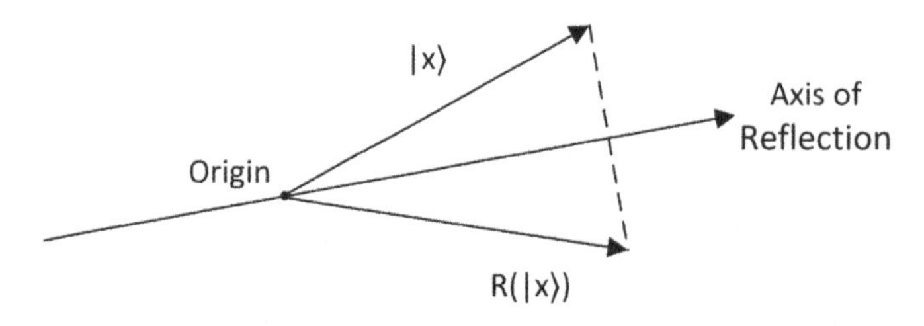

Figure 5.7 – A reflection transformation

Let's use the basis set E of $|e_1\rangle$ and $|e_2\rangle$ in the following diagram:

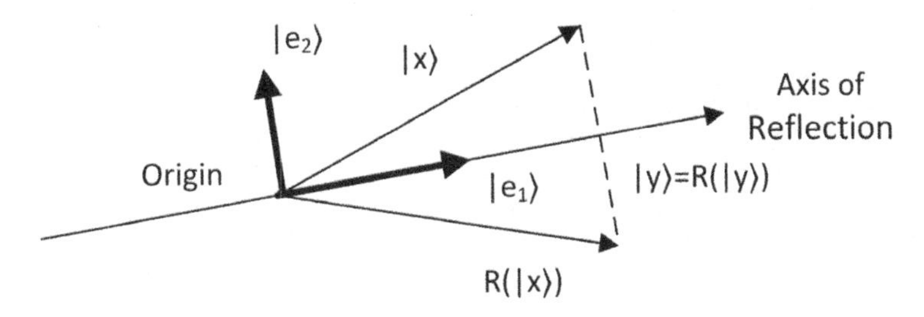

Figure 5.8 – A reflection transformation with basis vectors

Using this basis, we can see that:

$$|x\rangle = 2|e_1\rangle + |e_2\rangle$$

and

$$R(|x\rangle) = 2|e_1\rangle - |e_2\rangle$$

We can use the scalars of this linear combination as coordinates, as we learned in *Chapter 4, Vector Spaces*. This gives us the following:

$$|x\rangle = \begin{bmatrix} 2 \\ 1 \end{bmatrix}_E$$

$$R(|x\rangle) = \begin{bmatrix} 2 \\ -1 \end{bmatrix}_E$$

where E is our set of basis vectors.

So now, for our transformation R, we need to find a matrix A that represents it. Here is what I'm trying to say mathematically:

$$\text{Let } |x\rangle = \begin{bmatrix} 2 \\ 1 \end{bmatrix} \text{ then } R(|x\rangle) = \begin{bmatrix} 2 \\ -1 \end{bmatrix}$$

$$R(|x\rangle) = A|x\rangle = \begin{bmatrix} a_{11} & a_{12} \\ a_{21} & a_{22} \end{bmatrix}\begin{bmatrix} 2 \\ 1 \end{bmatrix} = \begin{bmatrix} 2a_{11} + 1a_{12} \\ 2a_{21} + 1a_{22} \end{bmatrix} = \begin{bmatrix} 2 \\ -1 \end{bmatrix}$$

If you look at the equation closely, you should be able to make out what the entries of the matrix should be. Here they are in all their glory!

$$R(|x\rangle) = A|x\rangle = \begin{bmatrix} 1 & 0 \\ 0 & -1 \end{bmatrix}\begin{bmatrix} 2 \\ 1 \end{bmatrix} = \begin{bmatrix} 2 \\ -1 \end{bmatrix}$$

This is great! We have now found a matrix that we can multiply any vector by in our 2D space to get its reflection using our basis vectors E. Let's do it for the vector $|y\rangle$ in this diagram:

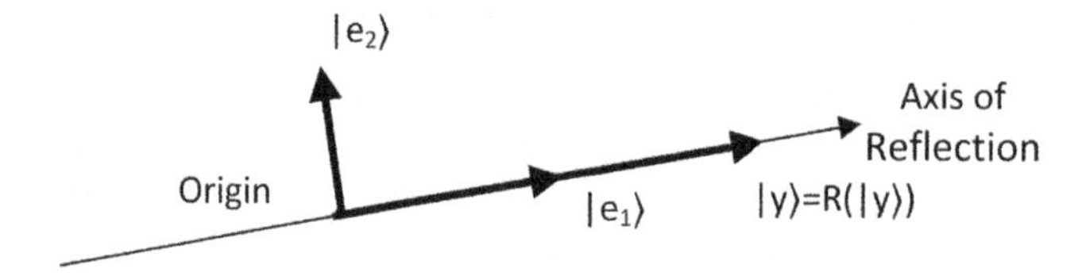

Figure 5.9 – A reflection of $|y\rangle$ with basis vectors

Since $|y\rangle$ is on the axis of reflection, it is its own reflection. In terms of our basis vectors E, $|y\rangle$ and its reflection have the following coordinates:

$$|y\rangle = R(|y\rangle) = \begin{bmatrix} 2 \\ 0 \end{bmatrix}_E$$

Alright, it's time for the moment of truth. Will our matrix A give us the right vector back? Let's check:

$$R(|y\rangle) = A|y\rangle = \begin{bmatrix} 1 & 0 \\ 0 & -1 \end{bmatrix}\begin{bmatrix} 2 \\ 0 \end{bmatrix} = \begin{bmatrix} 2 \\ 0 \end{bmatrix}$$

It does! I will go ahead and tell you that this matrix will work for any vector that is expressed in terms of our basis set E.

You should also be able to see that if we picked a different set of basis vectors, the matrix would be different as well. In our first example, we chose a matrix based implicitly on the canonical computational basis. This is called the **standard matrix** of the linear transformation. But if you change the basis, say to |+⟩ and |-⟩, the matrix will change as well. You can even get really complex and change the input and output bases of the transformation to affect a new matrix, but this is rarely done in practice. The takeaway point of this section is that a matrix can represent a linear transformation, but there are many matrices that can represent it based on the basis chosen. Finally, matrices that represent the same linear transformation are called **similar**.

Matrix multiplication and multiple transformations

The last superpower we will go over for matrices is the fact that not only can they represent transformations, but they can also represent multiple transformations through matrix multiplication!

Let's say that we wanted to do our reflection transformation twice. Intuitively, we should get back the same vector. Using matrix multiplication, we can show that algebraically:

$$R(|z\rangle) = A|z\rangle = \begin{bmatrix} 1 & 0 \\ 0 & -1 \end{bmatrix}\begin{bmatrix} a \\ b \end{bmatrix} = \begin{bmatrix} a \\ -b \end{bmatrix}$$

$$R(R(|z\rangle)) = A(A|z\rangle) = \begin{bmatrix} 1 & 0 \\ 0 & -1 \end{bmatrix}\begin{bmatrix} 1 & 0 \\ 0 & -1 \end{bmatrix}\begin{bmatrix} a \\ b \end{bmatrix} = \begin{bmatrix} 1 & 0 \\ 0 & -1 \end{bmatrix}\begin{bmatrix} a \\ -b \end{bmatrix} = \begin{bmatrix} a \\ b \end{bmatrix}$$

And there you have it - we have just proven that for any vector in $\mathbb{R}^2$, doing our reflection twice returns the same vector.

The commutator

You should remember from *Chapter 1, Superposition with Euclid*, that matrices do not, in general, commute. This also means that linear transformations do not commute in general. Physicists use something called the **commutator** to represent "how much" two transformations or matrices commute.

The commutator is defined to be:

$$[A, B] = AB - BA$$

for two n x n matrices. This holds for any matrices that represent a linear transformation. If the commutator is zero for two transformations, then they commute. If it is non-zero, the operators are said to be incompatible. In quantum mechanics, observables such as momentum are represented by linear transformations. All of this leads to the famous uncertainty principle that states that two observables that do not commute cannot be measured simultaneously.

Okay, let's move on to something a little less heady and talk about translations, rotations, and projections.

Transformations inspired by Euclid

In linear algebra, there are many "special" types of linear transformations that have names that connote concepts we have in our real world, such as reflections and projections. These concepts have been generalized to apply to all types of vectors, but the geometric description of them with Euclidean vectors gives us an idea as to why they work the way they do. This intuition can then be taken and applied to all types of vectors and vector spaces.

Translation

The first transform we will look at is translation. It transforms all vectors in a vector space by a displacement vector. More precisely:

$$T(|x\rangle) = |x\rangle + |d\rangle \quad \text{where} \quad |d\rangle \text{ is the vector of displacement}$$

In the following graph, the vector $|x\rangle$ is translated to the right by $|d\rangle$ to form $T(|x\rangle)$:

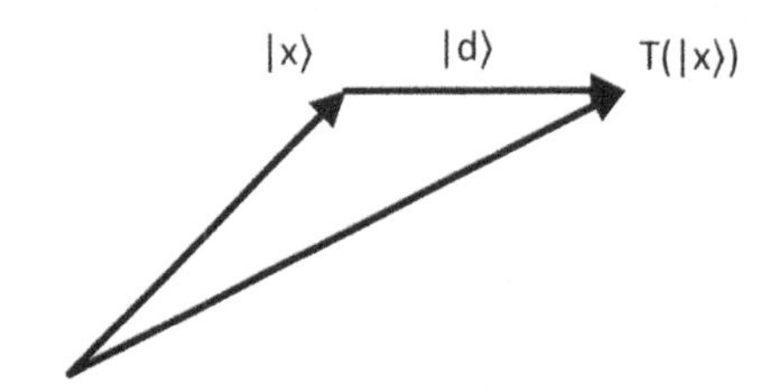

Figure 5.10 – A graphical depiction of translation

What's interesting about this type of translation is that it turns out to be non-linear! I will quickly show you.

Let's start with additivity and work it out algebraically. First, we will transform two vectors and add them together:

$$T(|u\rangle) = |u\rangle + |d\rangle \quad T(|v\rangle) = |v\rangle + |d\rangle$$

$$T(|u\rangle) + T(|v\rangle) = |u\rangle + |v\rangle + 2|d\rangle$$

This should equal the transformation of the two vectors added together:

$$T(|u\rangle + |v\rangle) = |u\rangle + |v\rangle + |d\rangle$$

Unfortunately, they are not equal. In other words:

$$T(|u\rangle) + T(|v\rangle) \neq T(|u\rangle + |v\rangle).$$

Another quick way to prove that a transformation is not linear is to show that the transform of the zero vector does not return the zero vector:

$$T(\mathbf{0}) \neq \mathbf{0}$$
$$T(\mathbf{0}) = |d\rangle$$

Finally, if we draw out the vectors, we can see that the transformation is not linear. The following diagram is a test for homogeneity and, as you can see, the transform $T(2|x\rangle)$ does not equal $2T(|x\rangle)$:

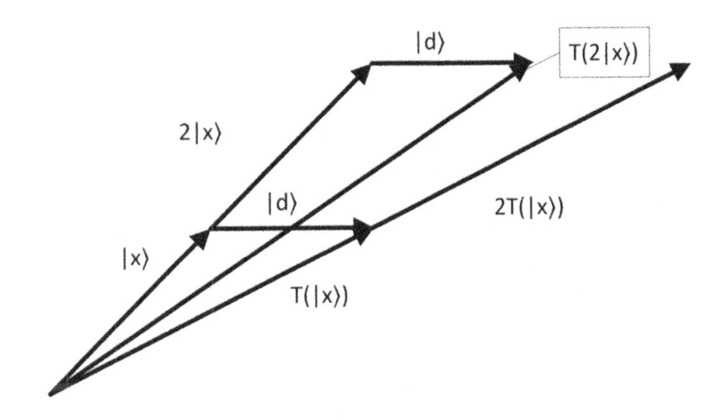

Figure 5.11 – A test for homogeneity

I could have left this transformation out of this section, but I thought it important to show you that even intuitive concepts such as translation can be non-linear. Now, let's look at the linear transformations of projection and rotation. No more trickery – all the remaining transformations are linear!

Rotation

Everyone has a concept of what a rotation is. We need to take that concept and express it mathematically. This section will rely a lot on trigonometry. If you need to brush up, please consult the *Appendix* chapter on trigonometry.

Okay, let's start with two-dimensional rotations. I will actually define them using the following graph:

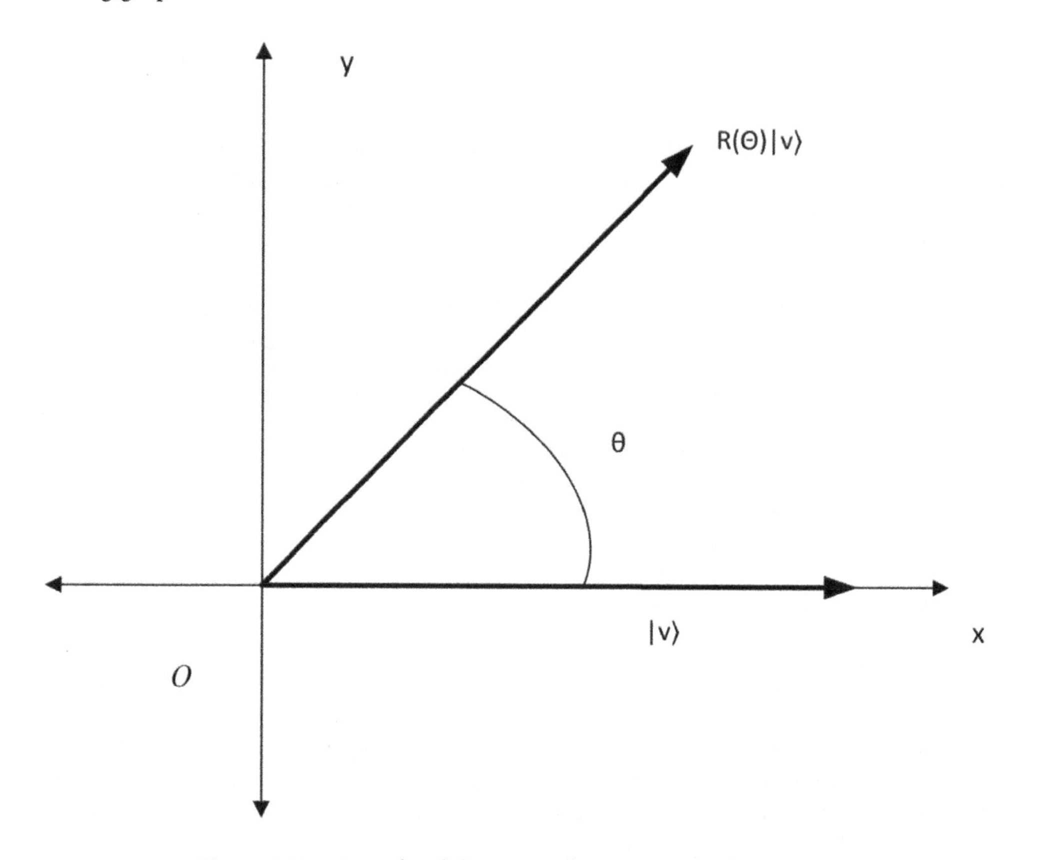

Figure 5.12 – A graphical depiction of a rotation transformation

Describing this in words, given a vector $|v\rangle$ and an angle θ, $R(\theta)$ will rotate $|v\rangle$ through an angle θ with respect to the x axis about the origin.

Let's see what this transformation does to our two computational basis vectors $|0\rangle$ and $|1\rangle$. First, $|0\rangle$ – if I rotate $|0\rangle$ by θ radians, what do I get? Let's look at a graph:

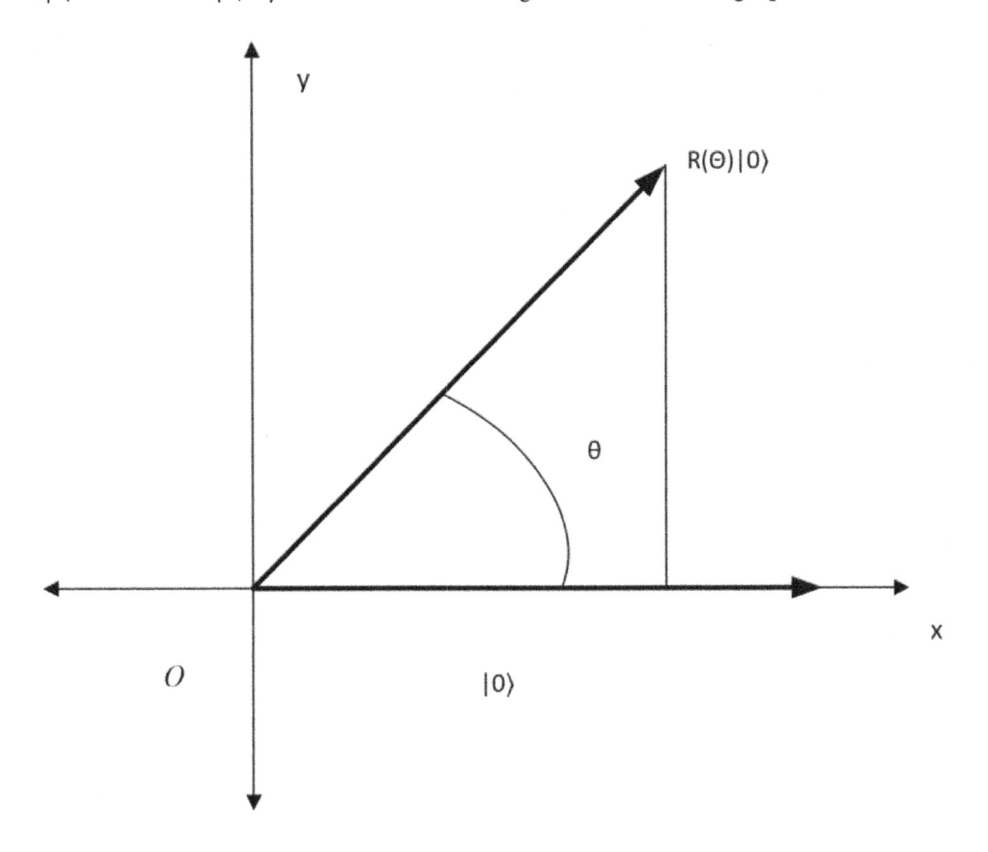

Figure 5.13 – The effect of rotation by Θ on $|0\rangle$

From the graph, we can tell that the new coordinates for $|0\rangle$ will be:

$$R(\theta)|0\rangle = \begin{bmatrix} \cos\theta \\ \sin\theta \end{bmatrix}$$

Now, on to $|1\rangle$. Let's look at its graph:

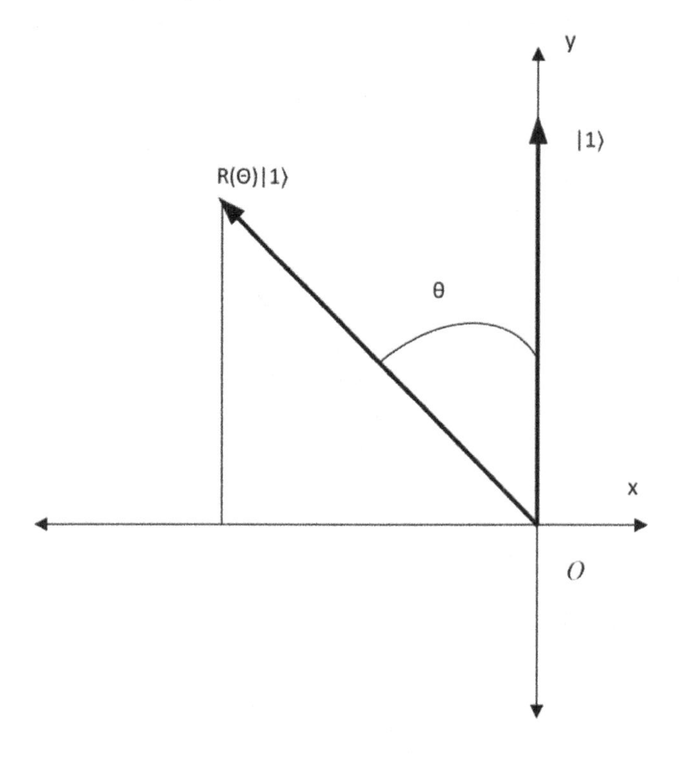

Figure 5.14 – The effect of rotation by Θ on $|1\rangle$

From the graph, we can tell that:

$$R(\theta)|1\rangle = \begin{bmatrix} -\sin\theta \\ \cos\theta \end{bmatrix}$$

I have now fully described the transformation both geometrically and through the basis vectors. What if I want to come up with a matrix for this transformation? Well, there is a theorem in linear algebra that if I give you the results of a transformation according to the computational basis, I can then compute the matrix according to this formula:

$$\begin{gathered} T(|x\rangle) = A|x\rangle \\ \text{if I know} \\ T(|0\rangle) \text{ and } T(|1\rangle) \\ \text{then} \\ A = \begin{bmatrix} T(|0\rangle) & T(|1\rangle) \end{bmatrix} \end{gathered} \tag{1}$$

In other words, I can use the results I found before of the transform's effect on the computational basis vectors. Taking those results as column vectors and putting them into a matrix gives me the standard matrix for the linear transformation! Without further ado, here is our result:

$$R(\theta)|0\rangle = \begin{bmatrix} \cos\theta \\ \sin\theta \end{bmatrix}$$

$$R(\theta)|1\rangle = \begin{bmatrix} -\sin\theta \\ \cos\theta \end{bmatrix}$$

$$R(\theta)(|x\rangle) = A|x\rangle$$

then

$$A = \begin{bmatrix} R(\theta)(|0\rangle) & R(\theta)(|1\rangle) \end{bmatrix} = \begin{bmatrix} \cos\theta & -\sin\theta \\ \sin\theta & \cos\theta \end{bmatrix}$$

Based on this, I can give you the result of a rotation for any vector in R^2:

$$|v\rangle = a|0\rangle + b|1\rangle = \begin{bmatrix} a \\ b \end{bmatrix}$$

$$R(\theta)|v\rangle = \begin{bmatrix} \cos\theta & -\sin\theta \\ \sin\theta & \cos\theta \end{bmatrix} \begin{bmatrix} a \\ b \end{bmatrix} = \begin{bmatrix} a\cos\theta - b\sin\theta \\ a\sin\theta + b\cos\theta \end{bmatrix}$$

Note that *Equation (1)* works for any finite amount of computational basis vectors as well:

$$T(|x\rangle) = A|x\rangle$$

if I know

$$T(|e_1\rangle), T(|e_2\rangle), \ldots, T(|e_n\rangle)$$

then

$$A = \begin{bmatrix} T(|e_1\rangle) \ldots & T(|e_n\rangle) \end{bmatrix}$$

where $|e_1\rangle, |e_2\rangle, \ldots, |e_n\rangle$ are the standard computational basis vectors.

Now, on to another intuitive transformation, projection.

Projection

Projection is another linear transformation that it is good to be acquainted with in quantum computing, since it is used heavily in the measurement of qubits. A good way to conceptually look at it is the way we think about projection in the everyday world. Let's take the process of taking a picture with a camera. When you do this, you are projecting a 3D world onto a 2D surface.

Also, if you were to take a picture of the picture, you would get the same picture. Doing a projection twice does not yield a different result.

In this figure, I am projecting a 3D cube onto a two-dimensional plane:

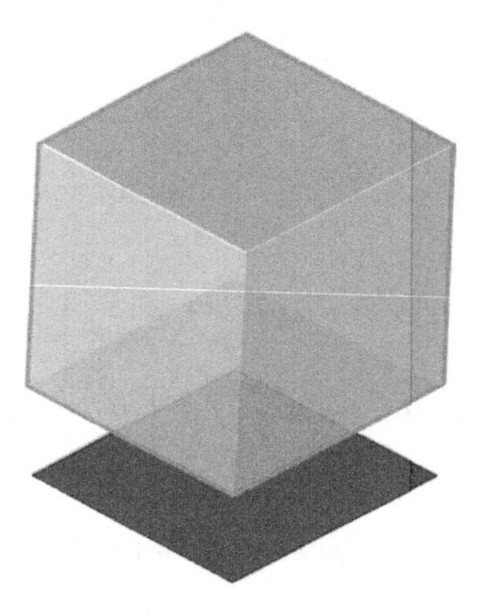

Figure 5.15 – A projection of a cube on a 2-D plane

In the following figure, I am projecting a two-dimensional circle onto a 1D line:

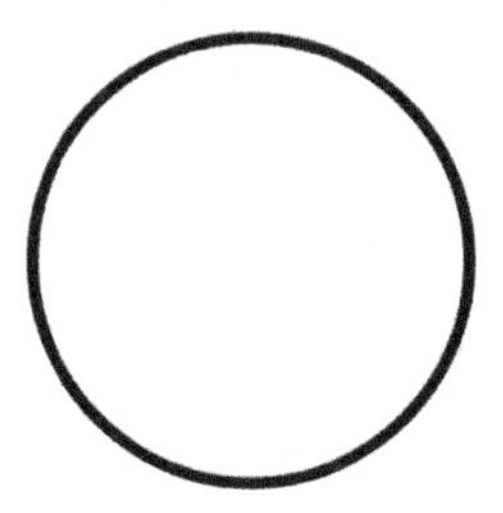

Figure 5.16 – A projection of a two-dimensional circle onto a one-dimensional line

If I project the line again, I will get the same line. It is this feature of projection that mathematicians have generalized to create a definition for projections. If you have a linear transformation P, then if the following condition holds, it is a projection:

$$P^2 = P \tag{2}$$

That's it! Not that bad, huh?

Okay, let's say that we want to come up with a projection matrix for the projection of the cube onto a plane in *Figure 5.15*. Let's say that the plane is the X-Y plane and all the points or vectors for the cube are in $\mathbb{R}^3$. So, we need to keep the X-Y coordinates but set the Z coordinates to zero. We need a matrix that does this:

$$P\begin{bmatrix} x \\ y \\ z \end{bmatrix} = \begin{bmatrix} x \\ y \\ 0 \end{bmatrix}$$

If we set P to the following matrix, that should do it:

$$P = \begin{bmatrix} 1 & 0 & 0 \\ 0 & 1 & 0 \\ 0 & 0 & 0 \end{bmatrix}$$

Let's see if *Equation (2)* holds:

$$P^2 = \begin{bmatrix} 1 & 0 & 0 \\ 0 & 1 & 0 \\ 0 & 0 & 0 \end{bmatrix} \cdot \begin{bmatrix} 1 & 0 & 0 \\ 0 & 1 & 0 \\ 0 & 0 & 0 \end{bmatrix} = \begin{bmatrix} 1 & 0 & 0 \\ 0 & 1 & 0 \\ 0 & 0 & 0 \end{bmatrix}$$

Indeed, it does. Now, you get to test it out.

Exercise two

What is the answer to this problem for a random three-dimensional vector?

$$P|v\rangle = \begin{bmatrix} 1 & 0 & 0 \\ 0 & 1 & 0 \\ 0 & 0 & 0 \end{bmatrix} \begin{bmatrix} 3 \\ -4 \\ 1 \end{bmatrix}$$

Now, on to a special type of linear transformation.

Linear operators

Linear operators are linear transformations that map vectors from and to the same vector space. Indeed, reflections, rotations, and projections are all linear operators. In quantum, we put a "hat" or caret on the top of the letter of the linear operator when we want to distinguish it from its representation as a matrix. For instance, all the following linear transformations are linear operators:

$$\hat{T}: \mathbb{R}^2 \rightarrow \mathbb{R}^2$$

$$T = \begin{bmatrix} 1 & 2 \\ 3 & 4 \end{bmatrix}$$

$$\hat{R}: \mathbb{R}^3 \rightarrow \mathbb{R}^3$$

$$R = \begin{bmatrix} 4 & 5 & 1 \\ 3 & 9 & 8 \\ 4 & 5 & 3 \end{bmatrix}$$

$$\hat{C}: \mathbb{C}^4 \rightarrow \mathbb{C}^4$$

$$\begin{bmatrix} i & 4i & 2 & -1 \\ -2i & 3 & 6 & 7 \\ 9 & 1 & 2 & 4 \\ 5 & 6 & 7 & 8 \end{bmatrix}$$

Most of the time, it is clear from the context that we are referring to a matrix or a linear operator, so the caret or "hat" is not used.

The following linear transformations are *not* linear operators:

$$T: \mathbb{R}^2 \rightarrow \mathbb{R}^3$$

$$\begin{bmatrix} 1 & 2 \\ 3 & 9 \\ 83 & 7 \end{bmatrix}$$

$$R: \mathbb{R}^6 \rightarrow \mathbb{R}^3$$

$$\begin{bmatrix} 8 & 9 & 8 & 5 & 4 & 6 \\ 1 & 5 & 3 & 4 & 9 & 4 \\ 6 & 3 & 3 & 5 & 4 & 6 \end{bmatrix}$$

You probably noticed that all linear operators are represented by square matrices. This leads to all types of special properties that they can have, such as determinants, eigenvalues, and invertibility. We will take up all these topics in later chapters, but I wanted to make sure that you knew this term. In quantum computing, you will rarely see the term *linear transformation*, but it is a common term used in mathematics. It will almost always be *linear operator* in quantum computing, and now you know that it is just a special type of linear transformation. Also, there is one other special type of linear transformation I would like to look at.

Linear functionals

A **linear functional** is a special case of a linear transformation that takes in a vector and spits out a scalar:

$$f : V \to F \text{ where } V \text{ is a vector space and } F \text{ is the field of scalars } \mathbb{R} \text{ or } \mathbb{C}$$

For instance, I could define a linear functional for every vector in R^2:

$$f\left(|v\rangle \right) = a + b \text{ where } |v\rangle = \begin{bmatrix} a \\ b \end{bmatrix}$$

So that:

$$f\left(\begin{bmatrix} 3 \\ 2 \end{bmatrix}\right) = 3 + 2 = 5$$

$$f\left(\begin{bmatrix} 5 \\ -2 \end{bmatrix}\right) = 5 - 2 = 3$$

There are many linear functionals that can be defined for a vector space. Here's another one:

$$g : \mathbb{R}^2 \to \mathbb{R}$$

$$g\left(|v\rangle \right) = 2a - 3b \text{ where } |v\rangle = \begin{bmatrix} a \\ b \end{bmatrix}$$

The set of all linear functionals that can be defined on a vector space actually form their own vector space called the dual vector space. This concept is important to fully define a bra in bra-ket notation. Please see the *Appendix* section on bra-ket notation if you are interested in more information.

A change of basis

We learned in *Chapter 4, Vector Spaces*, that a vector can have different coordinates depending on the basis that was chosen, but we didn't tell you how to go back and forth between bases. In this section, we will.

We want to come up with a matrix – let's call it B for a change of basis – that takes us from one basis to another. In other words, we want this mathematical formula to work:

$$B\begin{bmatrix} x_1 \\ \vdots \\ x_n \end{bmatrix}_C = \begin{bmatrix} y_1 \\ \vdots \\ y_n \end{bmatrix}_F$$

This matrix B will convert the coordinates of a vector according to a basis C to the coordinates for the vector in the basis F. Now, how do we find this matrix?

Let's look at an example. We will define the basis C as the computational basis and the basis F this way:

$$\text{Basis } C = \{|0\rangle, |1\rangle\} \quad \text{where } |0\rangle = \begin{bmatrix} 1 \\ 0 \end{bmatrix} \text{and } |1\rangle = \begin{bmatrix} 0 \\ 1 \end{bmatrix}$$

$$\text{Basis } F = \{|f_1\rangle, |f_2\rangle\} \text{ where } \quad |f_1\rangle = \begin{bmatrix} 1 \\ 1 \end{bmatrix} \text{ and } |f_2\rangle = \begin{bmatrix} -1 \\ 0 \end{bmatrix}$$

Now, let's look at a random vector, $|v\rangle$, defined in the computational basis C:

$$|v\rangle = 3|0\rangle + 4|1\rangle = 3\begin{bmatrix} 1 \\ 0 \end{bmatrix} + 4\begin{bmatrix} 0 \\ 1 \end{bmatrix} = \begin{bmatrix} 3 \\ 4 \end{bmatrix}_C \quad (3)$$

So, what we want to do is to find the coordinates of $|v\rangle$ in the basis F. In other words, we want to find the variables a and b in the following equation:

$$|v\rangle = a|f_1\rangle + b|f_2\rangle = \begin{bmatrix} a \\ b \end{bmatrix}_F$$

What would happen if we took our basis vectors in C and multiplied them by our change of basis matrix B? We would get our basis vectors in C expressed as coordinates in the basis F, as shown in the following equation:

$$B|0\rangle = |0\rangle_F$$
$$B|1\rangle = |1\rangle_F \quad (4)$$

Let's take our original *Equation (3)*, multiply it by our change of basis matrix *B*, and express it again based on our new-found knowledge from *Equation (4)*:

$$|v\rangle = 3|0\rangle + 4|1\rangle$$

$$B|v\rangle = 3(B|0\rangle) + 4(B|1\rangle) = 3|0\rangle_F + 4|1\rangle_F$$

We can express that very conveniently as matrix multiplication, like so:

$$B|v\rangle = \begin{bmatrix} |0\rangle_F & |1\rangle_F \\ & \end{bmatrix} \begin{bmatrix} 3 \\ 4 \end{bmatrix}_C \quad (5)$$

The next step is to find our basis vectors in *C*, $|0\rangle$ and $|1\rangle$, expressed as coordinates in *F*. To do this, we have to find them expressed as a linear combination of the vectors in the basis *F*! We'll start with $|0\rangle$:

$$|0\rangle = c|f_1\rangle + d|f_2\rangle$$

Let's work it all out:

$$|0\rangle = c|f_1\rangle + d|f_2\rangle = c\begin{bmatrix} 1 \\ 1 \end{bmatrix} + d\begin{bmatrix} -1 \\ 0 \end{bmatrix} = 0\begin{bmatrix} 1 \\ 1 \end{bmatrix} + (-1)\begin{bmatrix} -1 \\ 0 \end{bmatrix} = \begin{bmatrix} 1 \\ 0 \end{bmatrix}$$

$$|0\rangle_F = \begin{bmatrix} c \\ d \end{bmatrix} = \begin{bmatrix} 0 \\ -1 \end{bmatrix}$$

And here are the calculations to obtain the coordinates in *F* for $|1\rangle$:

$$|1\rangle = g|f_1\rangle + h|f_2\rangle = g\begin{bmatrix} 1 \\ 1 \end{bmatrix} + h\begin{bmatrix} -1 \\ 0 \end{bmatrix} = 1\begin{bmatrix} 1 \\ 1 \end{bmatrix} + 1\begin{bmatrix} -1 \\ 0 \end{bmatrix} = \begin{bmatrix} 0 \\ 1 \end{bmatrix}$$

$$|1\rangle_F = \begin{bmatrix} g \\ h \end{bmatrix} = \begin{bmatrix} 1 \\ 1 \end{bmatrix}$$

Now that we have that, let's plug these vectors back into *Equation (5)* and see what we get, using our change of basis matrix *B* on our random vector $|v\rangle$:

$$B|v\rangle = \begin{bmatrix} |0\rangle_F & |1\rangle_F \\ & \end{bmatrix} \begin{bmatrix} 3 \\ 4 \end{bmatrix}_C = \begin{bmatrix} 0 & 1 \\ -1 & 1 \end{bmatrix} \begin{bmatrix} 3 \\ 4 \end{bmatrix}_C = \begin{bmatrix} 4 \\ 1 \end{bmatrix}_F$$

So, there you have it – we have changed a vector expressed in C to a vector expressed in F using a matrix. Due to this, a change of basis is a linear transformation. With this change of basis matrix, we can change any vector expressed as coordinates in C to coordinates in F. But how do we know we're right? Well, the vector $|v\rangle$ should be equal as a linear combination in either basis, so:

$$|v\rangle = \begin{bmatrix} 3 \\ 4 \end{bmatrix}_C = \begin{bmatrix} 4 \\ 1 \end{bmatrix}_F$$

$$|v\rangle = 3|0\rangle + 4|1\rangle = 4|f_1\rangle + 1|f_2\rangle$$

$$|v\rangle = 3\begin{bmatrix} 1 \\ 0 \end{bmatrix} + 4\begin{bmatrix} 0 \\ 1 \end{bmatrix} = 4\begin{bmatrix} 1 \\ 1 \end{bmatrix} + 1\begin{bmatrix} -1 \\ 0 \end{bmatrix} = \begin{bmatrix} 3 \\ 4 \end{bmatrix}$$

Okay, now that we've worked out an example, I'll give you the general formula for transforming a basis, and hence, you will be able to transform the coordinates for every vector in one basis to another:

For two general bases

$$G = \{ |g_1\rangle \ldots |g_n\rangle \} \text{ and } H = \{ |h_1\rangle \ldots |h_n\rangle \}$$

The changes of basis matrix from G to H can be written as:

$$C_{G \to H} = \left[|g_1\rangle_H \ldots |g_n\rangle_H \right]$$

This is basically saying that you have to express the basis vectors in the input basis as coordinates in the output basis.

Summary

We have covered the breadth of linear transformations in this chapter. They are key to understanding the linear algebra that is pervasive in quantum computing. We've also seen how these transformations have been inspired by Euclidean geometry and considered special transformations such as linear operators and linear functionals. Finally, we saw how to do a change of basis, which is a linear transformation as well! Next up, we will go from real numbers into the field of complex numbers.

Answers to exercises

Exercise one

$$T : \mathbb{R}^4 \rightarrow \mathbb{R}^3$$

$$T(|x\rangle) = T\left(\begin{bmatrix} x \\ y \\ w \\ z \end{bmatrix}\right) = \begin{bmatrix} w \\ y \\ z \end{bmatrix} \text{ Yes, this is linear}$$

$$U : \mathbb{R}^2 \rightarrow \mathbb{R}^3$$

$$U(|x\rangle) = U\left(\begin{bmatrix} x \\ y \end{bmatrix}\right) = \begin{bmatrix} x + y \\ y \\ x^2 \end{bmatrix} \text{ No, this is non-linear}$$

$$Y : \mathbb{R}^3 \rightarrow \mathbb{R}^3$$

$$Y(|x\rangle) = Y\left(\begin{bmatrix} x \\ y \\ z \end{bmatrix}\right) = \begin{bmatrix} x + y \\ 2y \\ z + 2 \end{bmatrix} \text{ No, this is non-linear}$$

Exercise two

$$\begin{bmatrix} 3 \\ -4 \\ 0 \end{bmatrix}$$

Works cited

Geometry – What is the position of a unit cube so that its projection on the ground has maximal area? – *Mathematics Stack Exchange*

`https://math.stackexchange.com/questions/1894106/what-is-the-position-of-a-unit-cube-so-that-its-projection-on-the-ground-has-max`

Section 3: Adding Complexity

In this section, we add complex numbers to the mix. We use complex numbers to get into Eigenstuff. Then, we bring it all together in *Chapter 8*, *Our Space in the Universe*, and add on a layer of extra credit in *Chapter 9*, *Advanced Concepts*.

The following chapters are included in this section:

- *Chapter 6, Complex Numbers*
- *Chapter 7, Eigenstuff*
- *Chapter 8, Our Space in the Universe*
- *Chapter 9, Advanced Concepts*

6
Complex Numbers

"What is unpleasant here, and indeed directly to be objected to, is the use of complex numbers. Ψ is surely fundamentally a real function."

– Letter from Schrödinger to Lorentz. June 6, 1926.

Even the great physicist Erwin Schrödinger was perplexed by the occurrence of complex numbers in quantum mechanics. Yet, complex numbers have been found to be inherent to quantum mechanics and, hence, quantum computing. Up until now, we have concentrated on the real numbers to make concepts easier to get across. It is time now to cross the Rubicon and make our way into the complex plane. You have probably come across complex numbers before, but we will go into great depths about them in this chapter. I think it is unfortunate that René Descartes named i an imaginary number, as it makes it seem that this topic should be otherworldly. But so was the number *0* when it was introduced, and the same with negative numbers. No one gives them a second thought today and you should treat complex numbers the same way. They are just another set of numbers.

In this chapter, we are going to cover the following main topics:

- Three forms, one number
- Cartesian form
- Polar form
- The most beautiful equation in mathematics
- Exponential form
- Bloch sphere

Three forms, one number

There are three main ways of representing a complex number:

- Cartesian form (aka the general form)
- Polar form
- Exponential form

Each one has its advantages and disadvantages, depending on what we are trying to do. These will become evident as we go through them.

Definition of complex numbers

A complex number is a number that can be expressed in the following way:

$$z = a + bi \tag{1}$$

where a and b are real numbers and i is the **imaginary unit**. The imaginary unit is defined as:

$$i^2 = -1$$

It follows from this that there are two square roots of *-1*, *i* and *-i*.

The real part of a complex number is denoted by Re(z) and the imaginary part is denoted by Im(z). For our complex number, z, defined in *Equation (1)*, Re(z) = a and Im(z) = b. Two complex numbers, z and w, are equal if, and only if, Re(z) = Re(w) and Im(z) = Im(w).

It is important to remember that the set of real numbers $\mathbb{R}$ is a subset of $\mathbb{C}$. Hence, if Im(z) = 0 for a complex number z, then z is also a real number.

Let's quickly look at some examples of complex numbers:

$$z = 3 + 2i$$

$$w = -4.3 - \frac{2}{3}i$$

$$x = 6$$

$$y = 4i$$

Now, let's move on to describing the three forms of complex numbers.

Cartesian form

We used the Cartesian form to define a complex number. To see why it is called Cartesian, notice we can also use an ordered pair of real numbers to represent the complex number z. The first number of the ordered pair will be the real part of the complex number, and the second number will be the imaginary part:

$$z = a + bi = (a, b)$$

Given this, we can represent complex numbers on a Cartesian coordinate system since a and b are just real numbers. We will need to make a couple of modifications though.

We will replace the x axis with an axis for the real part of a complex number ($\mathrm{Re}(z)$), and the y axis with an axis for the imaginary part of a complex number ($\mathrm{Im}(z)$), like so:

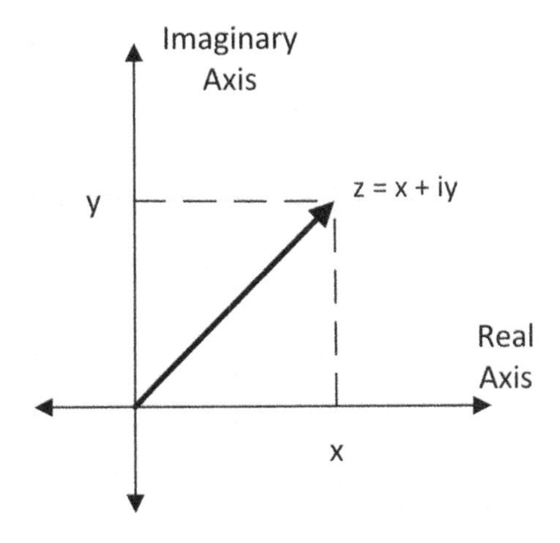

Figure 6.1 – The complex plane

This is called the complex plane. Here is an example involving actual complex numbers:

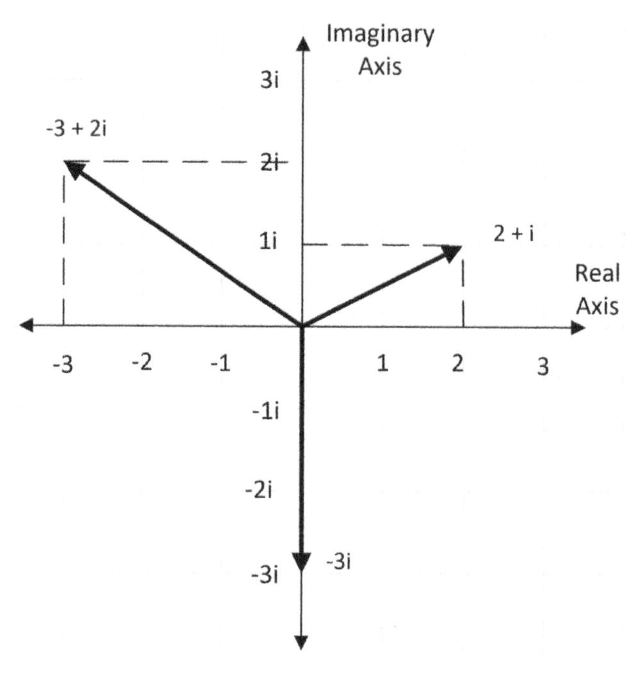

Figure 6.2 – Complex numbers on the complex plane

Keep this in mind as we go through the basic operations of complex numbers as we can think of them both algebraically and geometrically, as we did in *Chapter 2, Superposition with Euclid.*

Addition

Addition is rather easy for complex numbers; just add Re(z) and Im(z) of the two numbers together to get the sum. We will be using the following two complex numbers in our following definitions:

$$z_1 = a_1 + b_1 i$$

$$z_2 = a_2 + b_2 i$$

Here is the definition:

$$z_1 + z_2 = (a_1 + a_2) + (b_1 + b_2)i$$

Subtraction is defined as:

$$z_1 - z_2 = (a_1 - a_2) + (b_1 - b_2)i$$

Here are some examples:

$$(5 + 4i) + (7 + 2i)$$
$$12 + 6i$$

$$(-6 + 3i) + (5 + i)$$
$$-1 + 4i$$

$$i + (-5 + 6i)$$
$$-5 + 7i$$

Remember that you can view complex numbers as vectors on the complex plane, so addition, scalar multiplication, and subtraction can be viewed graphically as well, as in the following:

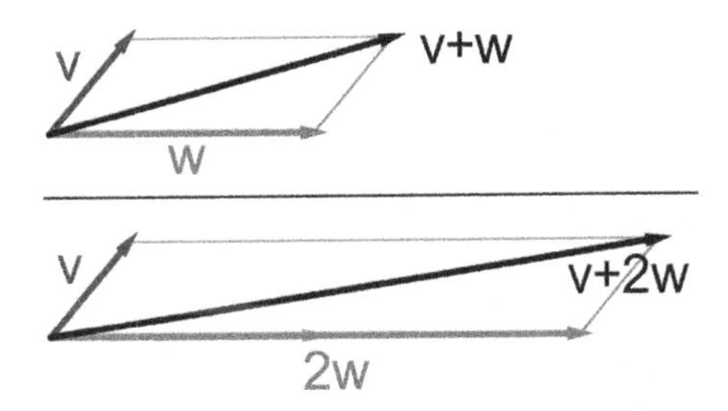

Figure 6.3 – Vector addition and scalar multiplication [1]

Alright, let's move on to multiplication.

Multiplication

I will show you another way to do complex multiplication later in this chapter, but you should know how to do it with the Cartesian form of a complex number. Hopefully, you remember the FOIL method from high school algebra. If you do, you can skip the next section. If not, here's a quick refresher:

FOIL method (optional)

The FOIL method is used to multiply two binomials together. It stands for:

- First terms
- Outer terms
- Inner terms
- Last terms

You add these all up, and there you are! This figure should jog your memory:

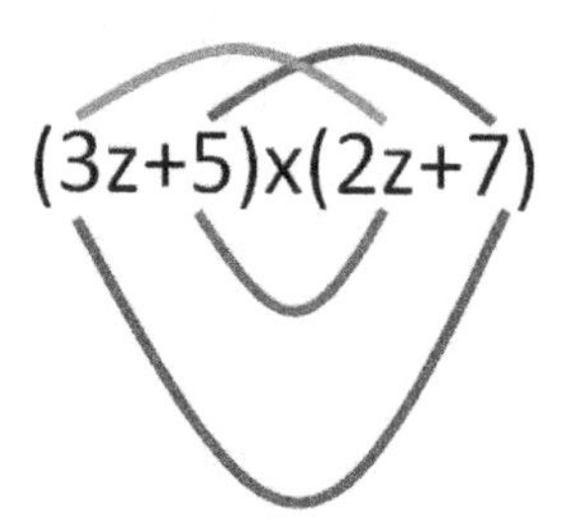

Firsts: 3z x 2z = 6z²

Outsides: 3z x 7 = 21z

Insides: 5 x 2z = 10z

Lasts: 5 x 7 = 35

$6z^2 + 21z + 10z + 35$

$= 6z^2 + 31z + 35$

Figure 6.4 – FOIL method illustrated [2]

Now that we've jogged your memory, on to the defintion of multiplication for complex numbers.

Definition

Here is the definition of the multiplication of two complex numbers in Cartesian form. It is:

$$
\begin{gathered}
z_1 \cdot z_2 = (a_1 + b_1 i)(a_2 + b_2 i) \\
a_1 a_2 + a_1 b_2 i + a_2 b_1 i + b_1 b_2 i^2 \\
a_1 a_2 + a_1 b_2 i + a_2 b_1 i - b_1 b_2 \\
(a_1 a_2 - b_1 b_2) + (a_1 b_2 + a_2 b_1) i
\end{gathered}
$$

As always, here is an example for your viewing pleasure:

$$
\begin{aligned}
(3 - 4i)(4 + 5i) &= 3 \cdot 4 + 3 \cdot 5i - 4i \cdot 4 - 4i \cdot 5i \quad \text{Use FOILMethod.} \\
&= 12 + 15i - 16i - 20i^2 \qquad \text{Substitute } i^2 = -1\,. \\
&= 12 + 15i - 16i - 20(-1) \\
&= 12 - i + 20 \\
&= 32 - i
\end{aligned}
$$

Now it's your turn.

Exercise 1

What is:

$$(2-3i)(5+i)$$

$$(1-i)(2+i)$$

$$(-2+3i)(-4-i)$$

Now for a new concept that doesn't exist for real numbers.

Complex conjugate

While the definition of the complex conjugate is very simple, store it somewhere safe in your brain as it will become very important to us as we move forward. The complex conjugate of a complex number $a + bi$ is $a - bi$. That's it! It is written as z^* for a complex number z. Here it is again, just to drill it into your skull :)

$$z = a + bi$$
$$z^* = a - bi$$

It is interesting to view complex conjugation on the complex plane as it is just a reflection of the real axis, as you can see in the following figure:

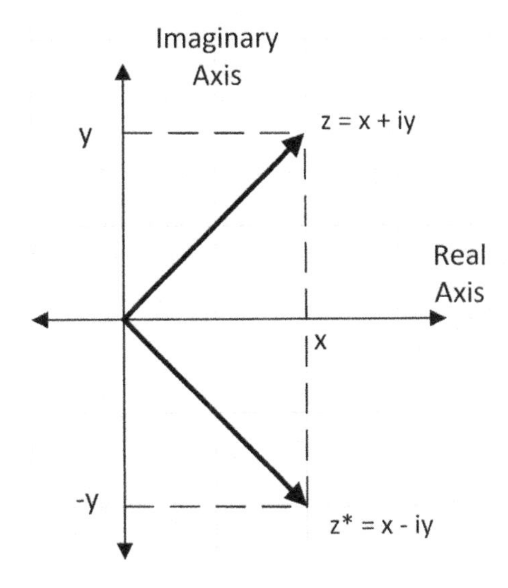

Figure 6.5 – Complex vector reflected on the real axis

Now we'll use the complex conjugate to get the absolute value of a complex number.

Absolute value or modulus

Again, simple, but very important. The absolute value or modulus of a complex number z, denoted $|z|$, is the square root of z multiplied by its conjugate, z^*:

$$|z| = \sqrt{z \cdot z^*}$$

It can also be defined thus for a complex number $z = x + iy$:

$$|z| = \sqrt{[Re(z)]^2 + [Im(z)]^2} = \sqrt{x^2 + y^2}$$

Here's an example:

$$|5 + 6i| = \sqrt{5^2 + 6^2}$$

$$= \sqrt{25 + 36}$$

$$= \sqrt{61}$$

Exercise 2

Compute the following absolute values:

$$|3 - 2i|$$

$$|i|$$

$$|6 + 3i|$$

Division

The way to compute the division of two complex numbers is unfortunately much harder than multiplication. However, there is a process named "rationalizing the denominator," which makes it easier.

Let's define our two complex numbers as:

$$\frac{c + di}{a + bi}$$

where both a and $b \neq 0$. First, we multiply the numerator and denominator by the complex conjugate of the denominator:

$$\frac{(c + di)}{(a + bi)} \cdot \frac{(a - bi)}{(a - bi)} = \frac{(c + di)(a - bi)}{(a + bi)(a - bi)}$$

Then we use the FOIL method:

$$= \frac{ca - cbi + adi - bdi^2}{a^2 - abi + abi - b^2i^2}$$

Finally, we substitute the terms that have i^2 with -1:

$$= \frac{ca - cbi + adi - bd(-1)}{a^2 - abi + abi - b^2(-1)}$$
$$= \frac{(ca + bd) + (ad - cb)i}{a^2 + b^2}$$

Hopefully, that wasn't too bad. Let's look at an example for this quotient:

$$\frac{(2 + 5i)}{(4 - i)}$$

From there, let's solve it together:

$$\frac{(2 + 5i)}{(4 - i)} \cdot \frac{(4 + i)}{(4 + i)} = \frac{8 + 2i + 20i + 5i^2}{16 + 4i - 4i - i^2}$$
$$= \frac{8 + 2i + 20i + 5(-1)}{16 + 4i - 4i - (-1)} \quad \text{Because } i^2 = -1$$
$$= \frac{3 + 22i}{17}$$
$$= \frac{3}{17} + \frac{22}{17}i \quad \text{Separate real and imaginary parts.}$$

As you can see, it's a little more than division in the real numbers.

Powers of *i*

I wanted to make sure that you could calculate the powers of *i* in your complex number toolkit. The positive powers of *i* follow this pattern:

$$i^0 = 1$$
$$i^1 = i$$
$$i^2 = -1$$
$$i^3 = i^2 \cdot i = -i$$
$$i^4 = i^3 \cdot i = 1$$
$$i^5 = i^4 \cdot i = i$$
$$i^6 = i^5 \cdot i = -1$$
$$i^7 = i^6 \cdot i = -i$$

The negative powers of i follow a very similar pattern:

$$i^0 = 1$$
$$i^{-1} = -i$$
$$i^{-2} = -1$$
$$i^{-3} = i$$
$$i^{-4} = 1$$
$$i^{-5} = -i$$
$$i^{-6} = -1$$
$$i^{-7} = i$$

Hopefully, you see the patterns here and can derive any power of *i*. Alright, off to the next form of complex numbers – polar!

Polar form

The polar form is based on polar coordinates, which you may or may not be used to. If not, the next section goes through these, otherwise, you can skip it. Also, the rest of the chapter is heavy on trigonometry and we use radians for all angles. If you require a quick refresher on these, please consult the *Appendix*.

Polar coordinates

Polar coordinates are another way of representing points in $\mathbb{R}^2$. We are very familiar with the Cartesian coordinate system and its points, such as (x,y). Now we will represent a point with two coordinates called r and θ. The following diagram is very helpful in terms of putting this all together:

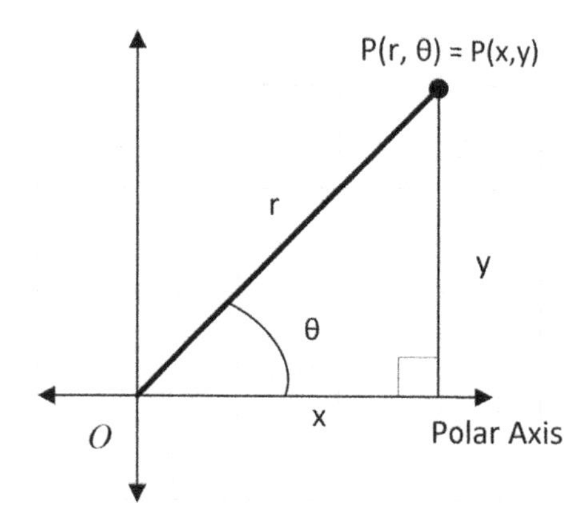

Figure 6.6 – Polar coordinates on a graph

As you can see, r is the hypotenuse of a right triangle with the other two sides being the Cartesian coordinates x and y. Because of this, it is easy to derive the equation to find r given the Cartesian coordinates using the Pythagorean theorem:

$$r^2 = x^2 + y^2 \tag{2}$$

We can use the trigonometric function tangent to derive θ:

$$tan\,\theta = \frac{y}{x}$$

$$\theta = arctan\left(\frac{y}{x}\right) = tan^{-1}\left(\frac{y}{x}\right) \tag{3}$$

Let's look at an example. Let's say we have the point (3,4), as shown in the following figure:

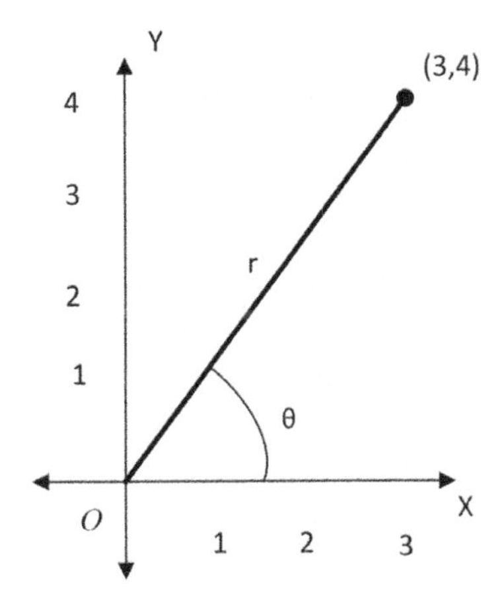

Figure 6.7 – The point (3,4)

Now we need to convert from Cartesian coordinates to polar coordinates using our formulas. So, using *Equation (2)*, r would be:

$$r^2 = x^2 + y^2$$
$$r^2 = 3^2 + 4^2$$
$$r^2 = 9 + 16$$
$$\sqrt{r^2} = \sqrt{25}$$
$$r = 5$$

Now that we've found r, we need θ. Again, we'll use our formula from *Equation (3)*:

$$\tan \theta = \frac{4}{3}$$

$$\theta = \tan^{-1}\left(\frac{4}{3}\right) \approx 0.9273 \text{ rad}$$

There you go! We have found that (3, 4) in Cartesian coordinates is (5, .9273) in polar coordinates. Now it is your turn.

Exercise 3

Convert the following into polar coordinates:

(-3,3)

(4, 1)

(-2, π)

Defining complex numbers in polar form

Since we can use Cartesian coordinates with complex numbers, we can also use polar coordinates with complex numbers. Here is *Figure 6.6* from the last section again, but this time using the complex plane:

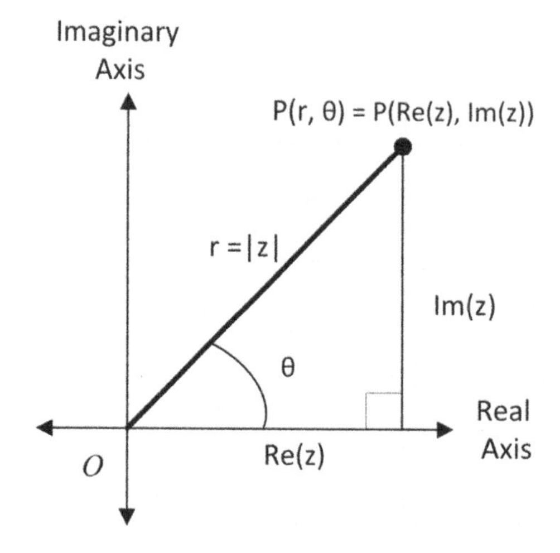

Figure 6.8 – Polar coordinates of the complex number z

Several things have changed in the graph. The real and imaginary axes have replaced the x and y axes, along with the x coordinate being Re(z) and the y coordinate being Im(z) now.

The r in the hypotenuse represents the modulus of our complex number, z. This makes sense if we remember that for a complex number, $z = a + bi$:

$$r = |z| = \sqrt{[Re(z)]^2 + [Im(z)]^2} = \sqrt{a^2 + b^2}.$$

The Greek letter Θ (pronounced "theta") is called the argument of the complex number z by mathematicians, but in quantum computing, it is often called the "phase." We denote this in math as follows:

$$arg(z) = \theta$$

From the graph, we can derive the following:

$$cos\,\theta = \frac{Re(z)}{|z|} \quad sin\,\theta = \frac{Im(z)}{|z|} \quad tan\,\theta = \frac{Im(z)}{Re(z)}$$

For $z = a + ib$, where $r = |z|$

$$a = r\,cos\,\theta \quad b = r\,sin\,\theta$$

Putting this all together, we can say that for a complex number, z, it can be represented in polar form as:

Polar Form of a Complex Number

$$\boxed{z = r(cos\,\theta + i\,sin\,\theta)}$$

Let's explore some more concepts of complex numbers in polar form.

Example

Okay, let's put this into an example. Let's find the polar form of $z = -4 + 4i$. First, we need to find the value of r:

$$r = \sqrt{a^2 + b^2}$$

$$r = \sqrt{(-4)^2 + 4^2}$$
$$r = \sqrt{16 + 16}$$
$$r = \sqrt{32} = 4\sqrt{2}$$

Now we have to find θ:

$$tan\,\theta = \frac{Im(z)}{Re(z)} = \frac{4}{-4} = -1$$

$$tan^{-1}(-1) = \frac{3\pi}{4}$$

$$\theta = \frac{3\pi}{4}$$

Now that we have *r* and θ, we can express *z* in polar form:

$$z = 4\sqrt{2}\left(cos\frac{3\pi}{4} + i\,sin\frac{3\pi}{4}\right)$$

Let's now explore some more concepts of complex numbers in polar form.

Multiplication and division in polar form

While the polar form is not recommended for addition and subtraction, it is recommended for multiplication and division. It is much easier to perform these operations, as you will quickly realize.

Given the two complex numbers below:

$$z_1 = r_1(cos\,\theta_1 + i\,sin\,\theta_1) \text{ and } z_2 = r_2(cos\,\theta_2 + i\,sin\,\theta_2)$$

The product of these two numbers is:

$$z_1 z_2 = r_1 r_2[cos(\theta_1 + \theta_2) + i\,sin(\theta_1 + \theta_2)] \tag{4}$$

Notice that all we had to do was add the angles and multiply the moduli. Pretty easy, right!?!

Dividing the two complex numbers is similarly easy. It is defined as:

$$\frac{z_1}{z_2} = \frac{r_1}{r_2}[cos(\theta_1 - \theta_2) + i\,sin(\theta_1 - \theta_2)], z_2 \neq 0$$

Here, all you have to do is subtract the angles and divide the moduli.

Example

Say we have two complex numbers:

$$z_1 = 2 + 2i$$

$$z_2 = \sqrt{3} - i$$

Converting to polar form, we have:

$$2 + 2i = 2\sqrt{2}\left(cos\frac{\pi}{4} + i\,sin\frac{\pi}{4}\right)$$
$$\sqrt{3} - i = 2\left[cos\left(-\frac{\pi}{6}\right) + i\,sin\left(-\frac{\pi}{6}\right)\right]$$

Applying our formula to the product of the two complex numbers in *Equation (4)*, we get the following for the product:

$$(2 + 2i)(\sqrt{3} - i) = 4\sqrt{2}\left[cos\left(\frac{\pi}{4} - \frac{\pi}{6}\right) + i\,sin\left(\frac{\pi}{4} - \frac{\pi}{6}\right)\right]$$
$$= 4\sqrt{2}\left(cos\frac{\pi}{12} + i\,sin\frac{\pi}{12}\right)$$

De Moivre's theorem

If we use *Equation (4)* to repeatedly multiply one complex number by itself, we get:

$$z = r(cos\,\theta + i\,sin\,\theta)$$
$$z^2 = r^2(cos\,2\,\theta + i\,sin\,2\,\theta)$$
$$z^3 = zz^2 = r^3(cos\,3\,\theta + i\,sin\,3\,\theta)$$

You should see a pattern whereby, in order to get a power of a complex number, we take the power of the modulus and multiply the angles by the power. This is known as **de Moivre's theorem**. It states that:

If $z = r(cos\,\theta + i\,sin\,\theta)$ and n is a positive integer, then

$$z^n = [r(cos\,\theta + i\,sin\,\theta)]^n = r^n(cos\,n\,\theta + i\,sin\,n\,\theta)$$

Make sure to tuck this away somewhere. Let's now move on to the most beautiful equation in mathematics!

The most beautiful equation in mathematics

In 1748, Leonhard Euler (pronounced "oy-lr") published his most famous formula, aptly called **Euler's Formula**:

$$e^{i\theta} = cos\,\theta + i\,sin\,\theta$$

This is true for any real number θ. Substituting θ = π into this equation gives the most beautiful equation in math, called **Euler's identity**:

$$e^{i\pi} + 1 = 0$$

It is hard to overstate the beauty of this equation. It combines in one equation what are arguably the most important symbols and operations in mathematics. Along with the operations of addition, multiplication, and exponentiation, you have 0, the additive identity, 1, the multiplicative identity, *i*, the imaginary unit, and two of the most important mathematical constants, *e* and π. This equation is also integral to quantum computing. You will see $e_i\theta$ all over the place in quantum computing, so you better get used to it!

So, how does this equation help us in quantum computing? You're about to find out, but first, we need to use it to express complex numbers in exponential form.

Exponential form

Complex numbers written in terms of $e^{i\theta}$ are said to be in the exponential form, as opposed to the polar or Cartesian form we have seen earlier. Using Euler's formula, we can express a complex number, *z*, as:

$$z = r(cos\,\theta + i\,sin\,\theta) = re^{i\theta}$$

So

$$z = re^{i\theta} \quad \text{in which } r = |z| \text{ and } \theta = arg(z)$$

As you can see, the exponential form is very close to polar form, but now you have θ in one place instead of two!

Exercise 4

Express the following complex numbers in exponential form:

$$z = 1 - i$$

$$z = 2 + 3i$$

$$z = -6$$

Conjugation

As we have seen, the conjugation of a complex number is represented as a reflection around the real axis. For complex numbers in exponential form, this means we just change the sign of the angle to get the complex conjugate:

$$\text{If} \quad z = r(cos\,\theta + i\,sin\,\theta) = re^{i\theta},$$

$$\text{then} \quad z^* = r(cos\,\theta - i\,sin\,\theta) = r(cos(-\theta) + i\,sin(-\theta)) = re^{-i\theta}$$

$$\text{Note that } cos\,\theta = cos(-\theta) \text{and } -sin\,\theta = sin(-\theta)$$

Multiplication

Multiplication and division are even easier in exponential form and are one of the reasons why it is so preferred to work with. We can take our steps for multiplication from the polar form and easily restate them in exponential form.

Given the two complex numbers below:

$$z_1 = r_1(\cos\theta_1 + i\sin\theta_1) \text{ and } z_2 = r_2(\cos\theta_2 + i\sin\theta_2)$$

In exponential form, they are represented as:

$$z_1 = r_1e^{i\theta_1} \text{ and } z_2 = r_2e^{i\theta_2}$$

The product of these two numbers in polar form is then:

$$z_1z_2 = r_1r_2[\cos(\theta_1 + \theta_2) + i\sin(\theta_1 + \theta_2)]$$

Their product in exponential form is then:

$$z_1z_2 = r_1r_2e^{i(\theta_1+\theta_2)}$$

Without going through all that, I will simply state that for division:

$$\frac{z_1}{z_2} = \frac{r_1}{r_2}e^{i(\theta_1-\theta_2)}, \quad z_2 \neq 0$$

Example

Let's reuse our example from the section on the polar form, but do it in the exponential form!

$$z_1 = 2 + 2i$$

$$z_2 = \sqrt{3} - i$$

Going from the Cartesian form to the polar form and then the exponential form, we get:

$$2 + 2i = 2\sqrt{2}\left(\cos\frac{\pi}{4} + i\sin\frac{\pi}{4}\right) = 2\sqrt{2}e^{\frac{i\pi}{4}}$$

$$\sqrt{3} - i = 2\left[\cos\left(-\frac{\pi}{6}\right) + i\sin\left(-\frac{\pi}{6}\right)\right] = 2e^{\frac{-i\pi}{6}}$$

Finally, using our definition of multiplication in the exponential form from before, we get:

$$(2 + 2i)(\sqrt{3} - i) = 4\sqrt{2}e^{\frac{i\pi}{4}}e^{\frac{-i\pi}{6}} = 4\sqrt{2}e^{\frac{i\pi}{12}}$$

Conjugate transpose of a matrix

Since we now have the definition of the complex conjugate of a number, I'd like to quickly go over the conjugate transpose of a matrix as we will use this later in the book. The conjugate transpose is exactly as it sounds. It combines the notions of complex conjugates and the transposition of a matrix into one operation. If you remember from *Chapter 2, The Matrix*, we defined the transpose to be:

$$\text{If } A = \begin{bmatrix} a_{11} & a_{12} & \cdots & a_{1n} \\ a_{21} & a_{22} & \cdots & a_{2n} \\ \vdots & \vdots & \ddots & \vdots \\ a_{m1} & a_{m2} & \cdots & a_{mn} \end{bmatrix}, \text{then } A^T = \begin{bmatrix} a_{11} & a_{21} & \cdots & a_{m1} \\ a_{12} & a_{22} & \cdots & a_{m2} \\ \vdots & \vdots & \ddots & \vdots \\ a_{1n} & a_{2n} & \cdots & a_{mn} \end{bmatrix}$$

This is where we essentially convert the rows into columns and the columns into rows.

The conjugate of a matrix is just the conjugation of every entry:

$$\text{If } A = \begin{bmatrix} a_{11} & a_{12} & \cdots & a_{1n} \\ a_{21} & a_{22} & \cdots & a_{2n} \\ \vdots & \vdots & \ddots & \vdots \\ a_{m1} & a_{m2} & \cdots & a_{mn} \end{bmatrix}, \text{then } A^* = \begin{bmatrix} a_{11}{}^* & a_{12}{}^* & \cdots & a_{1n}{}^* \\ a_{21}{}^* & a_{22}{}^* & \cdots & a_{2n}{}^* \\ \vdots & \vdots & \ddots & \vdots \\ a_{m1}{}^* & a_{m2}{}^* & \cdots & a_{mn}{}^* \end{bmatrix}$$

For example, if the matrix M equals

$$\begin{bmatrix} 1+2i & 4 \\ e^{\frac{i\pi}{2}} & -i \end{bmatrix},$$

then M^* equals

$$\begin{bmatrix} 1-2i & 4 \\ e^{\frac{-i\pi}{2}} & i \end{bmatrix}.$$

So here is the big payoff. The conjugate transpose of a matrix A is defined to be:

$$A^\dagger = (A^*)^T$$

The cross symbol at the top right of A is pronounced "dagger," and therefore when you hear "A dagger," the conjugate transpose of A is being referred to.

A quick example should get this all sorted. Let's use our matrix M from before. The conjugate transpose of M would be:

$$M = \begin{bmatrix} 1+2i & 4 \\ e^{\frac{i\pi}{2}} & -i \end{bmatrix} \qquad M^\dagger = \begin{bmatrix} 1-2i & e^{\frac{-i\pi}{2}} \\ 4 & i \end{bmatrix}$$

Please note that the conjugate transpose of a matrix also goes under the names **Hermitian conjugate** and **adjoint matrix**.

Okay, with all that squared away, we can get to some cool quantum computing stuff called the Bloch sphere!

Bloch sphere

This is the big payoff of the chapter, understanding the Bloch sphere! The Bloch sphere, named after Felix Bloch, is a way to visualize a single qubit. From *Chapter 1, Superposition with Euclid*, we know that a qubit can be represented in the following way:

$$|\psi\rangle = \alpha|0\rangle + \beta|1\rangle$$

We did not say this before, but now that we have introduced complex numbers, we can say that α and β are actually complex numbers.

Now we know that complex numbers take two real numbers to represent, it looks as though we will need four real numbers to characterize a qubit state. This is very hard to graph as we cannot visualize 4D space. Let's see whether we can decrease the number of real numbers required to represent a qubit state.

First, let's replace α and β with their exponential form to get:

$$|\psi\rangle = r_\alpha e^{i\phi_\alpha}|0\rangle + r_\beta e^{i\phi_\beta}|1\rangle$$

Now, let's rearrange the right side of the equation by taking out $e^{i\phi_\alpha}$ and distributing it. Notice that I need to subtract the phase of $e^{i\phi_\alpha}$ from the second term:

$$|\psi\rangle = e^{i\phi_\alpha}\left(r_\alpha|0\rangle + r_\beta e^{i(\phi_\beta - \phi_\alpha)}|1\rangle\right) \quad (5)$$

It ends up that **quantum mechanics** (**QM**) calls $e^{i\phi_\alpha}$ the "global phase" in *Equation (5)* and that it has no measurable effect on the qubit state. Therefore, QM lets us drop it. So now we have:

$$|\psi\rangle = r_\alpha|0\rangle + r_\beta e^{i(\phi_\beta - \phi_\alpha)}|1\rangle$$

The $(\varphi_\beta - \varphi_\alpha)$ in the second term is called the "relative phase." We will replace it with just one φ and also restrict it to go from 0 to 2π radians. Here is the new equation:

$$|\psi\rangle = r_\alpha|0\rangle + r_\beta e^{i\phi}|1\rangle$$

Okay, hopefully, you're keeping up. If you're keeping score at home, with all this mathematical contortion, we are now down from four dimensions to three dimensions to describe the qubit state. Can we get down to two dimensions? Let's try!

Another constraint we know is that α and β represent probability measurements and all probabilities must add up to one, so:

$$|\alpha|^2 + |\beta|^2 = 1$$

Because of this, the following must also be true:

$${r_\alpha}^2 + {r_\beta}^2 = 1$$

We can represent this on the first quarter of a unit circle like so:

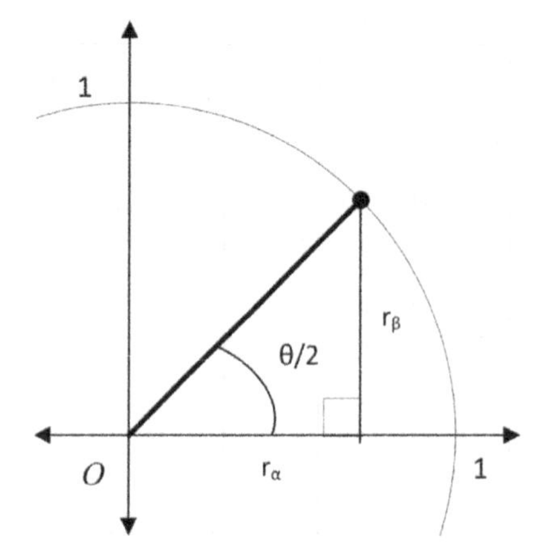

Figure 6.9 – Unit circle with a right triangle showing the parameters of our qubit

You may have noticed that we use θ/2 rather than just θ. The reason for this goes into very deep physics and math that are beyond the scope of this book. If you are very interested in why this is the case, please consult `https://physics.stackexchange.com/questions/174562/why-is-theta-over-2-used-for-a-bloch-sphere-instead-of-theta`.

Now we can represent our two coefficients for r as:

$$r_\alpha = cos\left(\frac{\theta}{2}\right)$$

$$r_\beta = sin\left(\frac{\theta}{2}\right)$$

$$0 \leq \theta < \pi$$

Okay, here's the big payoff!

$$|\psi\rangle = \cos\left(\tfrac{\theta}{2}\right)|0\rangle + \sin\left(\tfrac{\theta}{2}\right)e^{i\phi}|1\rangle$$

We're down to only two dimensions (θ and φ). Using these two angles, we can represent the qubit state on a unit sphere like so:

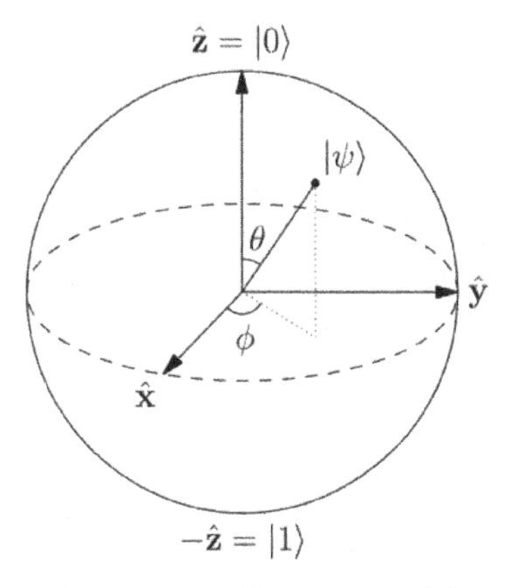

Figure 6.10 – Bloch sphere [3]

Θ is like the latitude on a map, but it goes from zero on the north pole, represented by the zero state, to π/2 on the equator and π on the south pole, which represents the one state. I was only able to show you how to get to the sphere mathematically, which is still a lot! For more information on why we need the Bloch sphere, please consult Packt's other great book, *Dancing with Qubits*, by Robert Sutor. For a cool, interactive visualization of the Bloch sphere, please go to `https://www.st-andrews.ac.uk/physics/quvis/simulations_html5/sims/blochsphere/blochsphere.html`.

Summary

Well, we've come a long way in a few pages. We've now seen how complex numbers can be expressed in the three forms of Cartesian, polar, and exponential. We've also seen which forms are better for different operations. You've learned how to multiply, divide, add, and subtract complex numbers. Finally, we've put it all together to mathematically derive the Bloch sphere. Bravo!

In the next chapter, we will explore the world of eigenvalues, eigenvectors, and all kinds of eigenstuff. Get ready!

Exercises

Exercise 1

$(2-3i)(5+i)=13-3i$

$(1-i)(2+i)=3-i$

$(-2+3i)(-4-i)=11-10i$

Exercise 2

$|3-2i|=\sqrt{13}$

$|i|=1$

$|6+3i|=3\sqrt{5}$

Exercise 3

(-3,3)

$\left(3\sqrt{2},\frac{3\pi}{4}\right)$

(4, 1)

$\left(\sqrt{17},.24498\right)$

(-2, π)

$\left(\sqrt{4+\pi^2},\pi-tan^{-1}\left(\frac{\pi}{2}\right)\right)$ or

$\left(-\sqrt{4+\pi^2},\pi-tan^{-1}\left(\frac{\pi}{2}\right)+2\pi\right)$

Exercise 4

$z=1-\mathrm{i}=\sqrt{2}e^{-i\pi/4}$

$z=2+3\mathrm{i}=\sqrt{13}e^{.9828i}$

$z=-6=6e^{i\pi}$

References

[1] – *File:Vector add scale.svg* – Wikimedia Commons (`https://commons.wikimedia.org/wiki/File:Vector_add_scale.svg`)

[2] – *MonkeyFaceFOILRule* – FOIL method – Wikipedia (`https://en.wikipedia.org/wiki/FOIL_method#/media/File:MonkeyFaceFOILRule.JPG`)

[3] – *File:Bloch Sphere.svg* – Wikimedia Commons (`https://commons.wikimedia.org/wiki/File:Bloch_Sphere.svg`)

Quote on Quantum physics needing complex numbers is from *2101.10873.pdf* (`https://arxiv.org/pdf/2101.10873.pdf`)

7
EigenStuff

Eigen (pronounced *EYE-GUN*) is a German prefix to words such as *eigentum* (property), *eigenschaft* (a feature or characteristic), and *eigensinn* (an idiosyncrasy). To sum up, we are looking for some values and vectors that are characteristic, idiosyncratic, and a property of something. What is that something? That something is our old friend the linear operator and its representations as square matrices. But before we get there, we'll need to look at some other concepts such as the matrix inverse and determinant. We'll wrap it all up with the trace of a matrix and some properties that the trace, determinant, and eigenvalues all share. These concepts will allow us to reach even further heights in the chapters that follow.

In this chapter, we are going to cover the following main topics:

- The inverse of a matrix
- Determinants
- Invertible matrix theorem
- Eigenvalues and eigenvectors
- Trace
- The special properties of eigenvalues

The inverse of a matrix

It would be nice to have a way to do algebra on matrices the way we do for simple algebraic expressions, like so:

$$3xy = 6$$

$$\text{divide both sides by } 3x$$

$$\frac{3xy}{3x} = \frac{6}{3x}$$

$$y = \frac{2}{x}$$

The inverse of a matrix provides us with a way to do this. It is very similar to the reciprocal for rational numbers. For rational numbers, the following is true:

$$x^{-1} = \frac{1}{x}$$

$$x \cdot x^{-1} = 1$$

In a similar way, the inverse of a matrix is defined to be a matrix that when multiplied by the original matrix, you get the identity matrix. Here it is mathematically:

$$A \cdot A^{-1} = I \quad \text{where } A^{-1} \text{ denotes the matrix inverse of } A$$

The matrix inverse can then be used when trying to algebraically modify a matrix equation. Let's say we are trying to find the vector $|x\rangle$ in the following equation:

$$A|x\rangle = |y\rangle$$

Since we now have a multiplicative inverse of a matrix, we can multiply both sides by it to get the following:

$$A^{-1}A|x\rangle = A^{-1}|y\rangle$$

$$I|x\rangle = A^{-1}|y\rangle$$

$$|x\rangle = A^{-1}|y\rangle$$

Please remember that matrix multiplication is not commutative, so if you left multiply a matrix on one side of an equation, you must left multiply on the other side. The same applies if you right multiply a matrix. We now have a way to find $|x\rangle$ by using the inverse of matrix A. But there is a catch – *not all matrices have an inverse.* Determinants will help us here though.

Determinants

Determinants *determine* whether a square matrix is invertible. This is a huge help to us, as we will see. In the literature, you will see either a function abbreviation for the determinant or vertical bars, like so:

$$\det(A) = |A|$$

The determinant is a function from $\mathbb{C}^{n \times n}$ to $\mathbb{C}$. In other words, it takes an n × n square matrix as input and spits out a scalar. For a 1 × 1 matrix, the determinant is just the number (easy enough). For a 2 × 2 matrix, this is the formula. You should probably just commit it to memory if you can:

$$\text{If } A = \begin{bmatrix} a & b \\ c & d \end{bmatrix},$$
$$\text{then } \det(A) = |A| = ad - bc$$

I will give you exercises at the end of this section to help with the memorization part, which will also give you a feel for the determinant itself.

There is a method for calculating determinants for bigger matrices, but it is rather involved, and once you've mastered 2 × 2 matrices, I would suggest using a matrix calculator. It's just like arithmetic; you should know how to do it for small numbers before using a calculator for everything else. Just so you get a glimpse of how it gets increasingly difficult to calculate determinants, here is the formula for a 3 × 3 matrix:

$$|A| = \begin{vmatrix} a & b & c \\ d & e & f \\ g & h & i \end{vmatrix} = a\begin{vmatrix} e & f \\ h & i \end{vmatrix} - b\begin{vmatrix} d & f \\ g & i \end{vmatrix} + c\begin{vmatrix} d & e \\ g & h \end{vmatrix}$$
$$= aei + bfg + cdh - ceg - bdi - afh.$$

For reference, a good matrix calculator is *WolframAlpha's* (`https://www.wolframalpha.com/calculators/determinant-calculator`), as shown here:

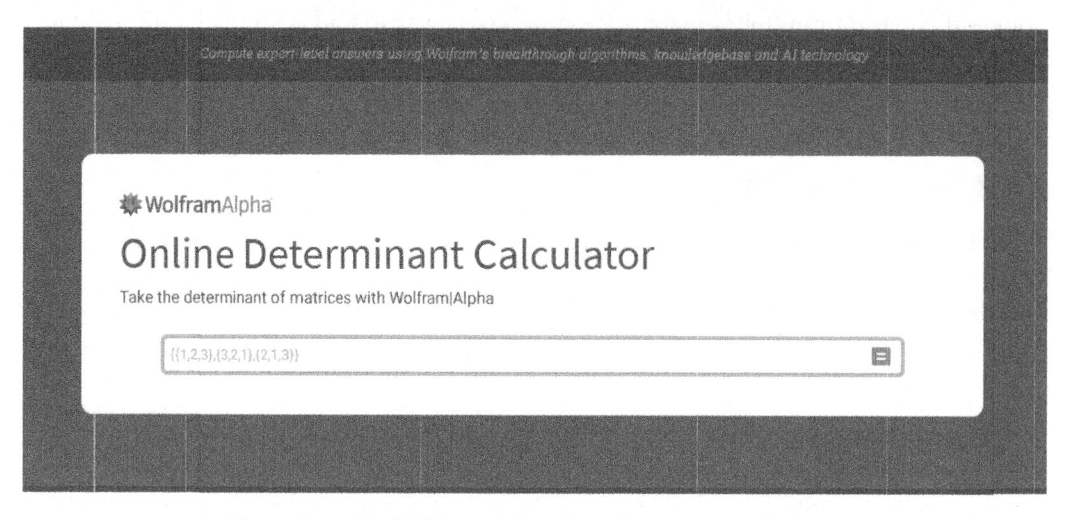

Figure 7.1 – The WolframAlpha online determinant calculator

Okay, let's see this in action and work through a quick example for determinants. What is the determinant of the following matrix?

$$\begin{bmatrix} -5 & -4 \\ -2 & -3 \end{bmatrix}$$

Using our formula, we can calculate it as follows:

$$\begin{vmatrix} -5 & -4 \\ -2 & -3 \end{vmatrix} = (-5)(-3) - (-4)(-2)$$
$$= 15 - 8$$
$$= 7$$

See? Easy-peasy!

So, at the start of this section, we said that the determinant helps to determine whether a matrix is invertible. How is that possible? I'll put this in a callout, as it is quite important:

> **Determinants and Invertibility**
>
> If the determinant of a matrix equals *zero*, it is not invertible. Otherwise, the matrix is invertible.

If a matrix is not invertible, we call it **singular** or **degenerate**. We will now use the determinant in the next sections on calculating the inverse of a matrix and calculating eigenvalues. But first, it leads us to a very important and useful theorem in linear algebra, the **invertible matrix theorem**.

Exercise one

What is the determinant of the following 2×2 matrices?

$$\begin{bmatrix} 1 & 2 \\ 3 & 4 \end{bmatrix}$$

$$\begin{bmatrix} 1 & 0 \\ 0 & 1 \end{bmatrix}$$

$$\begin{bmatrix} -1 & 3/2 \\ 1 & -1 \end{bmatrix}$$

The invertible matrix theorem

The invertible matrix theorem is a great result in linear algebra because based on the invertibility of a matrix, we can say a great many things about that matrix that are also true. Since we now know a quick way to determine the invertibility of a matrix through the computation of the determinant, we get all these other properties for free!

Here is the actual definition. Let A be a square $n \times n$ matrix. If A is invertible ($\det(A) \neq 0$), then the following properties follow:

- The column vectors of A are linearly independent.
- The column vectors of A form a basis for $\mathbb{C}^n$.
- The row vectors of A form a basis for $\mathbb{C}^n$ and are also linearly independent.
- The linear transformation mapping $|x\rangle$ to $A|x\rangle$ is a bijection from $\mathbb{C}^n$ to $\mathbb{C}^n$ (we studied bijections in *Chapter 3, Foundations*).

If the matrix A is not invertible ($\det(A) = 0$), then all the preceding properties are false.

While it may not seem that these are a lot of properties, there are many more I did not include because we didn't cover the concepts in this book. But even with the four properties listed here, it should show how powerful the determinant is when it is computed.

Calculating the inverse of a matrix

But wait – there's more! We can use the determinant of a matrix to calculate its inverse. Here is the formula to calculate the inverse of a 2 × 2 matrix:

$$A^{-1} = \begin{bmatrix} a & b \\ c & d \end{bmatrix}^{-1} = \frac{1}{\det(A)} \begin{bmatrix} d & -b \\ -c & a \end{bmatrix}$$

The first thing to notice is that the reciprocal of the determinant is used. Now, what if the determinant is zero? We can't divide by zero! So the formula is undefined, but remember that if the determinant is zero, it doesn't matter because the matrix is not invertible to begin with!

The other thing to notice is that a and d just switch position. Then, b and c are just multiplied by -1.

That was a lot of words! Let's look at an example with the following matrix:

$$D = \begin{bmatrix} 2 & 4 \\ -2 & 2 \end{bmatrix}$$

Let's calculate the determinant first:

$$\det(D) = |D| = 2 \cdot 2 - (-2)(4) = 4 + 8 = 12$$

Alright, let's use that to calculate the inverse:

$$D^{-1} = \frac{1}{12} \begin{bmatrix} 2 & -4 \\ 2 & 2 \end{bmatrix} = \begin{bmatrix} 1/6 & -1/3 \\ 1/6 & 1/6 \end{bmatrix}$$

And there you have it. Let's make sure that it is the inverse by using the definition we had for the inverse of a matrix:

$$DD^{-1} = \begin{bmatrix} 2 & 4 \\ -2 & 2 \end{bmatrix} \begin{bmatrix} 1/6 & -1/3 \\ 1/6 & 1/6 \end{bmatrix} = \begin{bmatrix} 1/3 + 2/3 & -2/3 + 2/3 \\ -1/3 + 1/3 & 2/3 + 1/3 \end{bmatrix} = \begin{bmatrix} 1 & 0 \\ 0 & 1 \end{bmatrix} = I$$

Now, let's do some exercises.

Exercise two

Calculate the inverse of the following matrices:

$$\begin{bmatrix} 1 & 2 \\ 3 & 4 \end{bmatrix}$$

$$\begin{bmatrix} 4 & 3 \\ 2 & -1 \end{bmatrix}$$

$$\begin{bmatrix} 1 & i \\ 2 & 2i \end{bmatrix}$$

You should note that in the last exercise, I included complex numbers in the matrix. You should get used to this now that we have covered complex numbers.

Now let's move on to the main feature, eigenvalues and eigenvectors!

Eigenvalues and eigenvectors

Ah, yes – we have arrived at the meat of this chapter, the eigen-stuff! We know that linear transformations are involved in this somehow, but how? Well, let's start with something we are familiar with, the reflection transformation we covered in *Chapter 5, Using Matrices to Transform Space*. Here is the diagram that describes it:

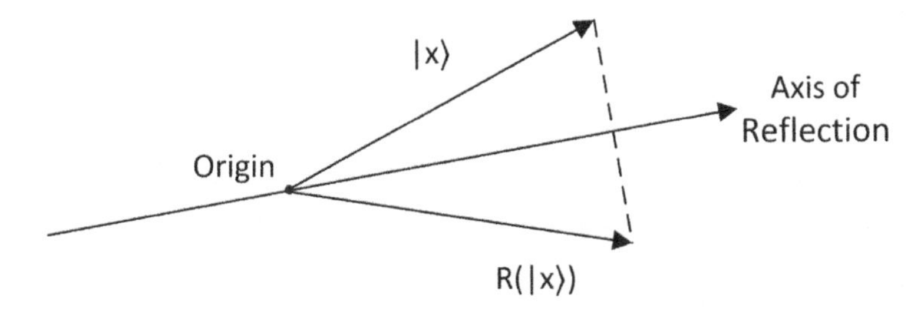

Figure 7.2 – Reflection transformation

Now, the special vectors we are looking for, called eigenvectors, are the ones that only get multiplied by a scalar when they are transformed. Here's a look at a number of vectors being reflected, with the reflections being the dashed lines. Stare at the diagram and see if you see any reflections that are the same as multiplying by a scalar:

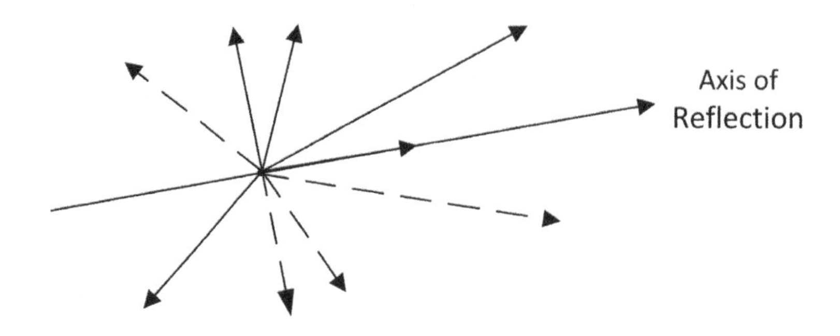

Figure 7.3 – Vectors and their reflections

Well, I don't want you to stare too long, so here is the first one:

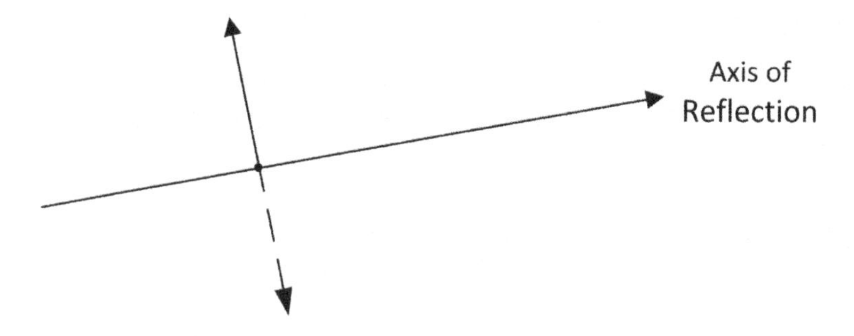

Figure 7.4 – The first eigenvector

Any vector along this line going through a reflection transformation is the same as being multiplied by the scalar -1. This scalar is called the eigenvalue.

Did you notice any other eigenvectors? This one is a little harder to tease out, but here it is:

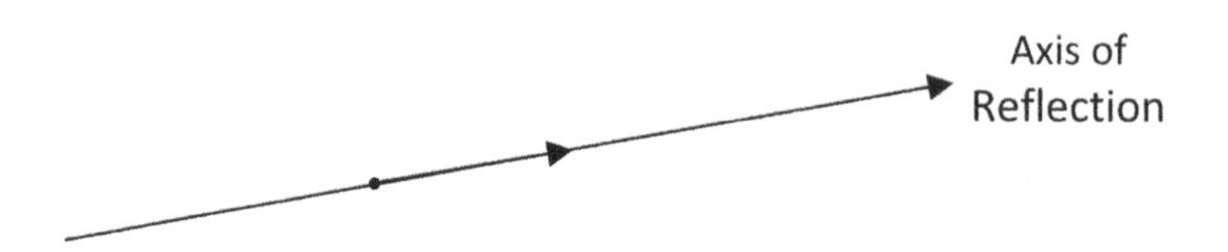

Figure 7.5 – The second eigenvector

If you remember in our discussion about reflection, we said that any vector that is on the axis of reflection will be reflected onto itself. In other words, it is multiplied by the scalar 1. The set of eigenvectors is any vector on this line, and the corresponding eigenvalue is 1.

So, there you have it – for reflection, an eigenvector is any vector perpendicular to the axis of reflection with an eigenvalue of -1 or any vector parallel to the axis of reflection with an eigenvalue of 1.

Definition

Let's get rigorous and define eigenvectors and eigenvalues. Given a linear operator $\hat{A}$ an eigenvector is a non-zero vector that when transformed by $\hat{A}$ is equal to being multiplied by a scalar λ (pronounced *lamb-da*). This can be written as the following:

$$\hat{A}|v\rangle = \lambda|v\rangle$$

This is also true for any matrix A that represents the linear operator, $\hat{A}$:

$$A\,|\,x\rangle = \lambda\,|\,x\rangle$$

where A is a matrix. The scalar λ is called an eigenvalue.

Now for some comic relief from all this rigor!

"What is a baby eigensheep?"

"A lamb, duh!"

Example with a matrix

Alright, let's see this eigen-stuff in action with an example matrix. Here's our chosen matrix:

$$B = \begin{bmatrix} 3 & 0 \\ 0 & 2 \end{bmatrix}$$

Let's see what it does to a variety of vectors:

$$\begin{bmatrix} 3 & 0 \\ 0 & 2 \end{bmatrix}\begin{bmatrix} 1 \\ 2 \end{bmatrix} = \begin{bmatrix} 3 \\ 4 \end{bmatrix}$$
$$\begin{bmatrix} 3 & 0 \\ 0 & 2 \end{bmatrix}\begin{bmatrix} -2 \\ 4 \end{bmatrix} = \begin{bmatrix} -6 \\ 8 \end{bmatrix}$$
$$\begin{bmatrix} 3 & 0 \\ 0 & 2 \end{bmatrix}\begin{bmatrix} 1 \\ 0 \end{bmatrix} = \begin{bmatrix} 3 \\ 0 \end{bmatrix}$$

Do any of the resultant vectors look like a scalar multiple of the original? If you said the last one, you are correct! Let's write it again with its eigenvalue, 3:

$$\begin{bmatrix} 3 & 0 \\ 0 & 2 \end{bmatrix}\begin{bmatrix} 1 \\ 0 \end{bmatrix} = \begin{bmatrix} 3 \\ 0 \end{bmatrix}$$
$$3\begin{bmatrix} 1 \\ 0 \end{bmatrix} = \begin{bmatrix} 3 \\ 0 \end{bmatrix}$$

You may have noticed that there is more than one eigenvector for an eigenvalue – in fact, there is a set. This leads us to the fact that an eigenvalue with its set of eigenvectors, plus the zero vector, constitutes a subspace of the overall vector space that a linear operator is transforming. For our matrix B, this happens to be a one-dimensional line through (1,0) if we are dealing with a real vector space. Let's see some more eigenvectors of the eigenvalue 3 to drive the point home:

$$\begin{bmatrix} 3 & 0 \\ 0 & 2 \end{bmatrix}\begin{bmatrix} -1 \\ 0 \end{bmatrix} = 3\begin{bmatrix} -1 \\ 0 \end{bmatrix} = \begin{bmatrix} -3 \\ 0 \end{bmatrix}$$
$$\begin{bmatrix} 3 & 0 \\ 0 & 2 \end{bmatrix}\begin{bmatrix} 2 \\ 0 \end{bmatrix} = 3\begin{bmatrix} 2 \\ 0 \end{bmatrix} = \begin{bmatrix} 6 \\ 0 \end{bmatrix}$$

This subspace of vectors is called the **eigenspace**. To define the eigenspace mathematically for the matrix B and the eigenvalue 3 in $\mathbb{R}^n$, it is as follows:

$$\left\{ \begin{bmatrix} x \\ 0 \end{bmatrix} : x \in \mathbb{R} \right\}$$

We can also just give the vector (1,0) as a basis vector. When doing this, the basis set of vectors is called the **eigenbasis**.

The characteristic equation

So, how do we find eigenvalues when given a square matrix? Well, let's start with what we are trying to find:

$$A|x\rangle = \lambda|x\rangle \tag{1}$$

We want to find λ. So, let's start with manipulating *Equation (1)* algebraically. Let's subtract $\lambda|x\rangle$ from both sides of the equation:

$$A|x\rangle - \lambda|x\rangle = \mathbf{0}$$

Okay, after doing that, let's take out the vector $|x\rangle$:

$$(A - \lambda)|x\rangle = \mathbf{0}$$

Hmm – this is not a valid equation because λ is a scalar and A is a matrix. What should we do? Let's multiply λ by the identity matrix so that we are subtracting two matrices:

$$(A - \lambda I)|x\rangle = \mathbf{0} \tag{2}$$

Okay, we now have one vector multiplied by the difference of two matrices on the left side of our equation and the zero vector on the other. It just so happens that a part of the invertible matrix theorem that I left unstated says that for *Equation (2)* to have a non-trivial solution (a solution other than the zero vector), the determinant of the matrix on the left side has to equal zero:

$$\det(A - \lambda I) = 0 \tag{3}$$

Equation (3) is called the **characteristic equation** of matrix A. If we can solve it, we will get all the eigenvalues of A! Let's see how we can do it with an example.

Example

Okay, let's use the matrix A, defined as follows:

$$A = \begin{bmatrix} 3 & -2 \\ 1 & -1 \end{bmatrix}$$

Now, let's solve its characteristic equation by first finding the matrix on the left in *Equation (3)*:

$$
\begin{aligned}
A - \lambda I &= \begin{bmatrix} 3 & -2 \\ 1 & -1 \end{bmatrix} - \lambda \begin{bmatrix} 1 & 0 \\ 0 & 1 \end{bmatrix} \\
&= \begin{bmatrix} 3 & -2 \\ 1 & -1 \end{bmatrix} - \begin{bmatrix} \lambda & 0 \\ 0 & \lambda \end{bmatrix} \\
&= \begin{bmatrix} 3-\lambda & -2 \\ 1 & -1-\lambda \end{bmatrix}
\end{aligned}
$$

Okay, now that we've found that, let's set the determinant of that matrix to 0 and solve for λ:

$$
\begin{aligned}
\det(A - \lambda I) &= (3-\lambda)(-1-\lambda) - (-2) = 0 \\
\det(A - \lambda I) &= -3 - 3\lambda + \lambda + \lambda^2 + 2 = 0 \\
\det(A - \lambda I) &= \lambda^2 - 2\lambda - 1 = 0
\end{aligned}
$$

The polynomial we have found in the preceding equation is called the **characteristic polynomial** of the matrix A for reference. There seems to be a lot of characteristics floating around, which gets back to our discussion of what *eigen* stands for in German! If we put the characteristic polynomial into the quadratic formula, we will find that our two eigenvalues for the matrix A are as follows:

$$\lambda = 1 \pm \sqrt{2}$$

And there you go - you now know how to find eigenvalues for 2×2 matrices.
What about bigger matrices? Well, the characteristic polynomials get harder and harder to solve symbolically, and we have to rely on computers to find them numerically. Once again, a handy matrix calculator comes to save the day. For getting eigenvalues and eigenvectors, I'd like to recommend *Symbolab* (`https://www.symbolab.com/`) as an online calculator as well (another tool in your toolbelt); a screenshot is shown in the following figure:

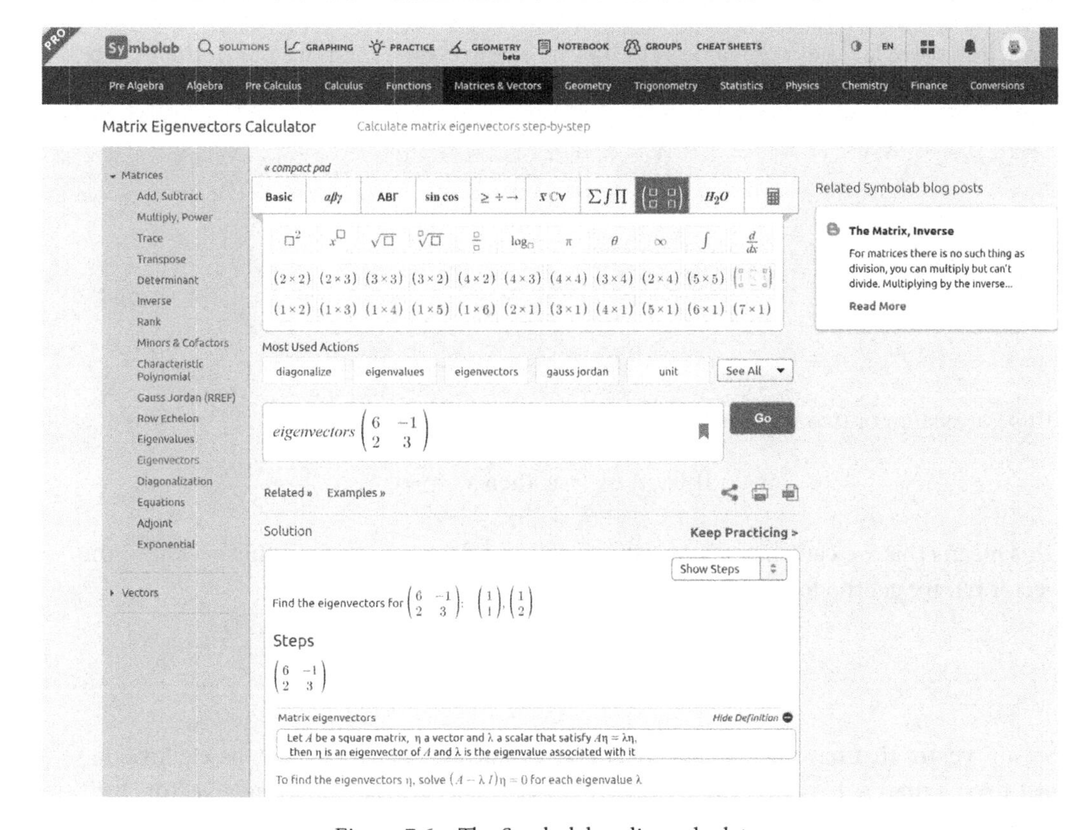

Figure 7.6 – The Symbolab online calculator

Now that we have found the eigenvalues, how do we find the eigenvectors?

Finding eigenvectors

In this section, we will answer that question. In a nutshell, we put the eigenvalues back into the definition and solve for the eigenvectors. Let's say I use the linear operator Y, which is a common gate in quantum computing. Here it is in the computational basis:

$$Y = \begin{bmatrix} 0 & -i \\ i & 0 \end{bmatrix}$$

I'll go ahead and tell you that the eigenvalues are 1 and -1. The definition for eigenvalues is as follows:

$$Y|x\rangle = \lambda|x\rangle$$

Now, let's solve for the eigenvalue 1, to find its eigenvectors. First, we put in the matrix and eigenvalue:

$$\begin{bmatrix} 0 & -i \\ i & 0 \end{bmatrix} \begin{bmatrix} x_1 \\ x_2 \end{bmatrix} = 1 \begin{bmatrix} x_1 \\ x_2 \end{bmatrix}$$

Multiplying this out gives us the following:

$$\begin{aligned} 0x_1 - ix_2 &= x_1 \\ ix_1 + 0x_2 &= x_2 \end{aligned}$$

This is a system of two linear equations, which is pretty simple to solve:

$$\text{If } 0x_1 - ix_2 = x_1 \text{ then } x_1 = -ix_2$$

This means that we can choose any value for x_2 and derive x_1. Putting this back into the vector $|x\rangle$, we get the following:

$$|x\rangle = \begin{bmatrix} -ix_2 \\ x_2 \end{bmatrix}$$

So, any vector that has the components of $|x\rangle$ will be an eigenvector for the eigenvalue of 1 for the matrix Y. Let's set one component to get a "representative eigenvector" for the eigenvalue:

$$\text{Let } x_2 = 1$$

$$\text{then } |x\rangle = \begin{bmatrix} -i \\ 1 \end{bmatrix}$$

To find the other set of eigenvectors for Y, you repeat the process with the other eigenvalue of -1.

Multiplicity

Sometimes, when we compute the characteristic polynomial for a matrix, it can output something like this:

$$(\lambda - 1)^2 = 0$$

In this case, the only eigenvalue is 1, but it is the root of the characteristic polynomial twice. Therefore, this particular eigenvalue has an algebraic multiplicity of 2. In general, the algebraic multiplicity of a particular eigenvalue is the number of times it appears as a root of the characteristic polynomial.

Trace

The trace of a matrix is very easy to compute, but it helps to have its value, as it has special properties. It is defined as the sum of the values on the main diagonal of a matrix. So, let's say we have the following:

$$A = \begin{bmatrix} a_{11} & a_{12} & a_{13} \\ a_{21} & a_{22} & a_{23} \\ a_{31} & a_{32} & a_{33} \end{bmatrix} = \begin{bmatrix} 1 & 3 & 5 \\ i & -i & 6 \\ 7 & \sqrt{2} & 3 \end{bmatrix}$$

The trace of A will then be the following:

$$\text{tr}(\mathbf{A}) = \sum_{i=1}^{3} a_{ii} = a_{11} + a_{22} + a_{33} = 1 - i + 3 = 4 - i$$

In general, the definition of the trace for an $n \times n$ square matrix is as follows:

$$\text{tr}(\mathbf{A}) = \sum_{i=1}^{n} a_{ii} = a_{11} + a_{22} + \ldots + a_{nn}$$

The trace of a linear operator is the same for any matrix that represents it.

The special properties of eigenvalues

Now that we have gone through the concept of the trace and determinant, let's quickly look at some cool properties they have in relation to eigenvalues.

It happens that the sum of the eigenvalues of a matrix equals the trace of the matrix:

$$\text{tr}(\mathbf{A}) = \sum_{i} \lambda_i$$

Additionally, the product of all the eigenvalues of a matrix is equal to the determinant:

$$\det(\mathbf{A}) = \lambda_1 \cdot \lambda_2 \cdot \ldots \cdot \lambda_n = \prod_{i} \lambda_i$$

I encourage you to go back to the matrices we have used as examples with their eigenvalues and prove to yourself that this is indeed true!

Summary

We have covered quite a bit of ground here in relatively few pages. We saw how to calculate the determinant and how it plays a role in finding the inverse of matrices. Many types of eigen-stuff were discussed, and ways to find them were also given. Finally, we saw how the trace is calculated and how it relates to everything else we covered in this chapter. These concepts will equip you to do serious linear algebra and quantum computing going forward. In the next chapter, we will bring everything together that we learned in the last seven chapters to explore the space where qubits live!

Answers to exercises

Exercise one

$$\det\begin{bmatrix} 1 & 2 \\ 3 & 4 \end{bmatrix} = -2$$

$$\det\begin{bmatrix} 1 & 0 \\ 0 & 1 \end{bmatrix} = 1$$

$$\det\begin{bmatrix} -1 & 3/2 \\ 1 & -1 \end{bmatrix} = -1/2$$

Exercise two

$$\begin{bmatrix} 1 & 2 \\ 3 & 4 \end{bmatrix}^{-1} = \frac{1}{2}\begin{bmatrix} -4 & 2 \\ 3 & -1 \end{bmatrix}$$

$$\begin{bmatrix} 4 & 3 \\ 2 & -1 \end{bmatrix}^{-1} = \frac{1}{10}\begin{bmatrix} 1 & 3 \\ 2 & -4 \end{bmatrix}$$

$$\begin{bmatrix} 1 & i \\ 2 & 2i \end{bmatrix} \text{ this matrix is not invertible as the det} = 0$$

8
Our Space in the Universe

Our space in the universe is called a **Hilbert space**. A Hilbert space is a type of vector space that has certain properties – properties that we will develop in this chapter. The most important will be defining an inner product. Once this is done, we will be able to measure distance and angles between vectors in an n-dimensional complex space. We will also be able to measure the length of vectors in these spaces. Later in the chapter, we will look at putting these Hilbert spaces together into even bigger Hilbert spaces through the tensor product!

The other main topic of this chapter is **linear operators**. We will go through many types, showing the distinct properties of each.

In this chapter, we are going to cover the following main topics:

- The inner product
- Orthonormality
- The outer product
- Operators
- Types of operators
- Tensor products

The inner product

An inner product can actually be any function that follows a few properties, but we are going to zero in on one definition of the inner product that we will use in quantum computing. Here it is:

$$\left(|x\rangle, |y\rangle \right) \equiv \left(\begin{bmatrix} x_1 \\ \vdots \\ x_n \end{bmatrix}, \begin{bmatrix} y_1 \\ \vdots \\ y_n \end{bmatrix} \right) = \sum_{i=1}^{n} x_i^* y_i = x_1^* y_1 + \cdots + x_n^* y_n$$

Mathematicians use the preceding notation for the inner product, but Dirac defined it with a bra and ket, calling it a bracket:

$$\langle x | y \rangle \equiv \left(|x\rangle, |y\rangle \right)$$

Now, if we define a bra to be the conjugate transpose of its corresponding ket, so that if:

$$|x\rangle = \begin{bmatrix} x_1 \\ \vdots \\ x_n \end{bmatrix}$$

Then, $\langle x|$ is now:

$$\begin{bmatrix} x_1^* & \cdots & x_n^* \end{bmatrix}$$

We can then define a bracket as just matrix multiplication!

$$\langle x | y \rangle = \begin{bmatrix} x_1^* & \cdots & x_n^* \end{bmatrix} \begin{bmatrix} y_1 \\ \vdots \\ y_n \end{bmatrix} = \sum_{i=1}^{n} x_i^* y_i = x_1^* y_1 + \cdots + x_n^* y_n$$

Pretty cool, eh? That is one of the reasons why bra-ket notation is so convenient! You should notice something else too. The bra $\langle x|$ is a linear functional. It will take any vector $|y\rangle$ and give you a scalar according to the inner product formula!

Let's look at an example. Let's say $|x\rangle$ and $|y\rangle$ are defined this way:

$$|x\rangle = \begin{bmatrix} 3 \\ i \\ 2i \end{bmatrix} \quad |y\rangle = \begin{bmatrix} 2 \\ -i \\ 3 \end{bmatrix}$$

Then, the bras are going to be:

$$\langle x| = \begin{bmatrix} 3 & -i & -2i \end{bmatrix} \quad \langle y| = \begin{bmatrix} 2 & i & 3 \end{bmatrix}$$

Now, let's calculate the inner product $\langle y|x\rangle$:

$$\langle y|x\rangle = \begin{bmatrix} 2 & i & 3 \end{bmatrix} \cdot \begin{bmatrix} 3 \\ i \\ 2i \end{bmatrix} = \begin{cases} 2\cdot 3 + i\cdot i + 3\cdot 2i \\ 6 - 1 + 6i \\ 5 + 6i \end{cases}$$

Let's reverse the inner product and calculate $\langle x|y\rangle$:

$$\langle x|y\rangle = \begin{bmatrix} 3 & -i & -2i \end{bmatrix} \cdot \begin{bmatrix} 2 \\ -i \\ 3 \end{bmatrix} = \begin{cases} 3\cdot 2 + -i\cdot -i + -2i\cdot 3 \\ 6 - 1 - 6i \\ 5 - 6i \end{cases}$$

You might have noticed that the preceding answers are complex conjugates of each other; that's because:

$$\langle x|y\rangle = \langle y|x\rangle^* .$$

Since bras are linear functionals, brackets are left-distributive (α and β are scalars):

$$\langle y| \left(\alpha|x\rangle + \beta|z\rangle \right) = \alpha\langle y|x\rangle + \beta\langle y|z\rangle \tag{1}$$

And they are also right-distributive:

$$\left(\alpha\langle x| + \beta\langle z| \right)|y\rangle = \alpha\langle x|y\rangle + \beta\langle z|y\rangle$$

Let's test the left-distributive property in an example where we evaluate both sides of *Equation (1)* separately. We will define all the variables first. Our two scalars will be $\alpha = 3$ and $\beta = 2$. Our vectors are defined as follows:

$$|x\rangle = \begin{bmatrix} 2 \\ 1 \\ 0 \end{bmatrix} \quad |z\rangle = \begin{bmatrix} 6 \\ 4 \\ 2 \end{bmatrix} \quad \langle y| = \begin{bmatrix} 1 & 3 & 4 \end{bmatrix}$$

Alright, here we go. Let's evaluate the left side of *Equation (1)* first:

$$\langle y|\left(\alpha|x\rangle + \beta|z\rangle\right) = \begin{bmatrix} 1 & 3 & 4 \end{bmatrix}\left(3\begin{bmatrix} 2 \\ 1 \\ 0 \end{bmatrix} + 2\begin{bmatrix} 6 \\ 4 \\ 2 \end{bmatrix}\right)$$

$$\begin{bmatrix} 1 & 3 & 4 \end{bmatrix}\left(\begin{bmatrix} 6 \\ 3 \\ 0 \end{bmatrix} + \begin{bmatrix} 12 \\ 8 \\ 4 \end{bmatrix}\right)$$

$$\begin{bmatrix} 1 & 3 & 4 \end{bmatrix}\begin{bmatrix} 18 \\ 11 \\ 4 \end{bmatrix} = 18 + 33 + 16 = 67$$

And now for the right side of *Equation (1)*:

$$\alpha\langle y|x\rangle + \beta\langle y|z\rangle = 3\left(\begin{bmatrix} 1 & 3 & 4 \end{bmatrix}\begin{bmatrix} 2 \\ 1 \\ 0 \end{bmatrix}\right) + 2\left(\begin{bmatrix} 1 & 3 & 4 \end{bmatrix}\begin{bmatrix} 6 \\ 4 \\ 2 \end{bmatrix}\right)$$

$$3(2 + 3 + 0) + 2(6 + 12 + 8)$$

$$3 \cdot 5 + 2 \cdot 26$$

$$15 + 52 = 67$$

Luckily, our answers match. Whew! I didn't want to have to go back and do that again! Let's move on to see how the inner product can be used in vector spaces.

Orthonormality

In this section, we will look at the concepts of the norm and orthogonality to come up with orthonormality.

The norm

We can define a metric on our vector spaces called the **norm** and denote it this way, $\|x\|$, where x is the vector on which the norm is being measured. In two- and three-dimensional Euclidean spaces, it is often called the **length** of a vector, but in higher dimensions, we use the term *norm*. It gives us a way to measure vectors.

We define the norm using our inner product from the previous section, like so:

$$\| x \| = \sqrt{\langle x | x \rangle}$$

As always, let's look at an example. What is the norm of the vector $|x\rangle$ here?

$$|x\rangle = \begin{bmatrix} 2 \\ 3 \\ 4 \end{bmatrix}$$

Well, let's work it out:

$$\begin{aligned} \| x \| &= \sqrt{\langle x | x \rangle} \\ &= \sqrt{\begin{bmatrix} 2 & 3 & 4 \end{bmatrix} \begin{bmatrix} 2 \\ 3 \\ 4 \end{bmatrix}} \\ &= \sqrt{4+9+16} = \sqrt{29} \end{aligned}$$

As you can see, the norm, $\|x\|$, of $|x\rangle$ is the square root of 29.

Normalization and unit vectors

Oftentimes, especially in quantum computing, we will want to represent our vectors by something called a **unit vector**. The word *unit* refers to the fact that the norm of these vectors is one. How do we achieve this? Well, we divide a vector by its norm in a process called **normalization**.

A unit vector $|u\rangle$ is found for the vector $|v\rangle$ using the following formula:

$$|u\rangle = \frac{|v\rangle}{\| v \|}$$

What is the unit vector for the vector $|x\rangle$ from the previous section? Well, we just divide $|x\rangle$ by its norm to obtain:

$$|u\rangle = \frac{|x\rangle}{\| x \|} = \frac{1}{\sqrt{29}} \begin{bmatrix} 2 \\ 3 \\ 4 \end{bmatrix} = \begin{bmatrix} 2/\sqrt{29} \\ 3/\sqrt{29} \\ 4/\sqrt{29} \end{bmatrix}$$

You should take the norm of $|u\rangle$ to ensure that it is indeed 1.

Orthogonality

Orthogonal may not be a word you've seen before, but I bet you are familiar with the word **perpendicular**. For instance, the two vectors in the following graph are perpendicular to each other:

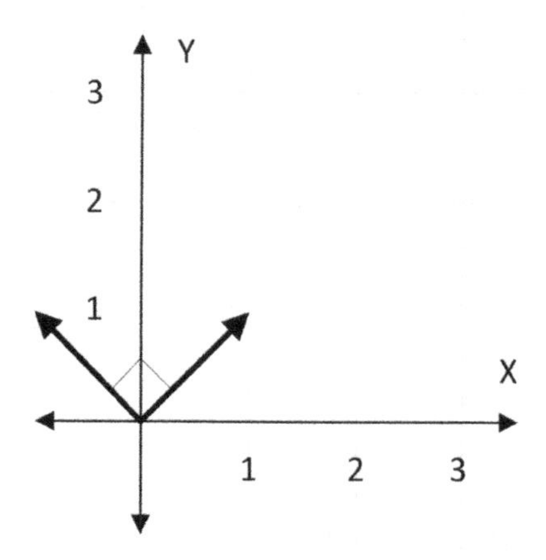

Figure 8.1 – Two perpendicular vectors

Orthogonality is taking the concept of vectors being perpendicular to each other to higher dimensions of $\mathbb{C}^n$.

The inner product is helpful in this regard as well. If two vectors are orthogonal to each other, their inner product will be 0. Let's see whether the two vectors in *Figure 8.1* are orthogonal according to the inner product:

$$\begin{bmatrix} -1 & 1 \end{bmatrix}\begin{bmatrix} 1 \\ 1 \end{bmatrix} = -1\cdot 1 + 1\cdot 1 = -1 + 1 = 0$$

Indeed, they are orthogonal, as their inner product is zero. Are the next two vectors orthogonal?

$$|x\rangle = \begin{bmatrix} 2 \\ 1 \\ 0 \end{bmatrix} \quad |z\rangle = \begin{bmatrix} 6 \\ 4 \\ 2 \end{bmatrix}$$

Well, let's see!

$$\langle x|z\rangle = \begin{bmatrix} 2 & 1 & 0 \end{bmatrix}\begin{bmatrix} 6 \\ 4 \\ 2 \end{bmatrix}$$

$$\langle x|z\rangle = 12 + 4 + 0 = 16$$

So, no – they're not! Now, it's your turn.

Exercise one

Determine whether each pair of vectors is orthogonal using the inner product:

1. $\begin{bmatrix} 1 \\ 2 \\ 3 \end{bmatrix}, \begin{bmatrix} 2 \\ -2 \\ -1 \end{bmatrix}$
2. $\begin{bmatrix} 0 \\ 1 \\ 2 \\ -1 \end{bmatrix}, \begin{bmatrix} -1 \\ 1 \\ 2 \\ 5 \end{bmatrix}$

Orthonormal vectors

So, if we combine the concept of orthogonality and normalization, we get orthonormal vectors. These are very important in the study of vector spaces and quantum computing.

Let's see whether we can come up with some orthonormal vectors in $\mathbb{C}^2$. We'll try the computational basis vectors:

$$|0\rangle = \begin{bmatrix} 1 \\ 0 \end{bmatrix} \quad |1\rangle = \begin{bmatrix} 0 \\ 1 \end{bmatrix}$$

Are they orthogonal?

$$\langle 1|0\rangle = \begin{bmatrix} 0 & 1 \end{bmatrix} \begin{bmatrix} 1 \\ 0 \end{bmatrix} = 0 \cdot 1 + 1 \cdot 0 = 0$$

Well, their inner product is zero, so they are orthogonal! Are they unit vectors?

$$\| 0 \|^2 = \langle 0|0\rangle = \begin{bmatrix} 1 & 0 \end{bmatrix} \begin{bmatrix} 1 \\ 0 \end{bmatrix} = 1 \cdot 1 + 0 \cdot 0 = 1$$

$$\| 1 \|^2 = \langle 1|1\rangle = \begin{bmatrix} 0 & 1 \end{bmatrix} \begin{bmatrix} 0 \\ 1 \end{bmatrix} = 0 \cdot 0 + 1 \cdot 1 = 1$$

Yes! The square root of their own inner product (aka their norm) is one. So $|0\rangle$ and $|1\rangle$ are orthonormal (*hint*: that's one of the reasons they were chosen as the canonical 0 and 1 vectors).

Let's look at another set of vectors that are important in quantum computing. They are labeled with plus and minus signs:

$$|+\rangle = \begin{bmatrix} \frac{1}{\sqrt{2}} \\ \frac{1}{\sqrt{2}} \end{bmatrix} \quad |-\rangle = \begin{bmatrix} \frac{1}{\sqrt{2}} \\ -\frac{1}{\sqrt{2}} \end{bmatrix}$$

What do you think? Orthonormal? Or just normal? How do we know? That's right – we use our formulas. Let's see whether they are orthogonal first:

$$\langle - | + \rangle = \begin{bmatrix} \frac{1}{\sqrt{2}} & -\frac{1}{\sqrt{2}} \end{bmatrix} \begin{bmatrix} \frac{1}{\sqrt{2}} \\ \frac{1}{\sqrt{2}} \end{bmatrix} = \begin{cases} \frac{1}{\sqrt{2}} \cdot \frac{1}{\sqrt{2}} + \left(-\frac{1}{\sqrt{2}} \right) \cdot \left(\frac{1}{\sqrt{2}} \right) \\ \frac{1}{2} - \frac{1}{2} = 0 \end{cases}$$

Okay, so their inner product is zero. Therefore, they are orthogonal. Are they normal or abnormal (he-he)?

$$\| + \|^2 = \langle + | + \rangle = \begin{bmatrix} \frac{1}{\sqrt{2}} & \frac{1}{\sqrt{2}} \end{bmatrix} \begin{bmatrix} \frac{1}{\sqrt{2}} \\ \frac{1}{\sqrt{2}} \end{bmatrix} = \frac{1}{2} + \frac{1}{2} = 1$$

$$\| - \|^2 = \langle - | - \rangle = \begin{bmatrix} \frac{1}{\sqrt{2}} & -\frac{1}{\sqrt{2}} \end{bmatrix} \begin{bmatrix} \frac{1}{\sqrt{2}} \\ -\frac{1}{\sqrt{2}} \end{bmatrix} = \frac{1}{2} + \frac{1}{2} = 1$$

Looks like they're normal! So $|+\rangle$ and $|-\rangle$ are orthonormal vectors.

The Kronecker delta function

Almost all vectors in quantum computing are unit vectors and many are orthogonal to each other. When you are dealing with orthonormal vectors and doing computations, it is convenient to use the **Kronecker delta function**. It is defined thusly:

$$\delta_{ij} = \begin{cases} 0 & \text{if } i \neq j \\ 1 & \text{if } i = j \end{cases}$$

So, when you see that symbol, if the indices of i and j are equal, then it is one. Otherwise, it is zero.

A good example is the identity matrix. If you replace the entries with the Kronecker delta function, you will get ones on the main diagonal when the indices of the entries equal each other:

$$I = \begin{bmatrix} \delta_{11} & \delta_{12} \\ \delta_{21} & \delta_{22} \end{bmatrix} = \begin{bmatrix} 1 & 0 \\ 0 & 1 \end{bmatrix}$$

More importantly, when you have a basic set of orthonormal vectors, the inner product of a vector with itself is one because they are normalized. The inner product is zero when you use any other basis vector because they are orthogonal. For instance, take the basis set B with orthonormal vectors:

$$B = \left\{ |0\rangle, |1\rangle, |2\rangle, ..., |n\rangle \right\},$$

We can then use the Kronecker delta function to succinctly represent their inner product:

$$\langle i \mid j \rangle = \delta_{ij}$$

I want you to know this function because you will see it in quantum computing literature, and it helps us in succinctly describing computations.

The outer product

The outer product is interesting because it is again the matrix multiplication of two vectors, but this time, the vectors are reversed, producing a matrix. The inner product uses the matrix multiplication of two vectors to get a scalar. The outer product uses two vectors to produce a matrix. Formally, the outer product is defined to be:

$$|u\rangle\langle v| = \begin{bmatrix} u_1 v_1{}^* & u_1 v_2{}^* & \cdots & u_1 v_n{}^* \\ u_2 v_1{}^* & u_2 v_2{}^* & \cdots & u_2 v_n{}^* \\ \vdots & \vdots & \ddots & \vdots \\ u_m v_1{}^* & u_m v_2{}^* & \cdots & u_m v_n{}^* \end{bmatrix}$$

If you remember matrix multiplication from *Chapter 2*, *The Matrix*, we have the following situation when multiplying an m × n matrix and an n × p matrix. They produce an m × p matrix, as shown in the following diagram. Since we are dealing with vectors, we have an m × 1 matrix and a 1 × p matrix:

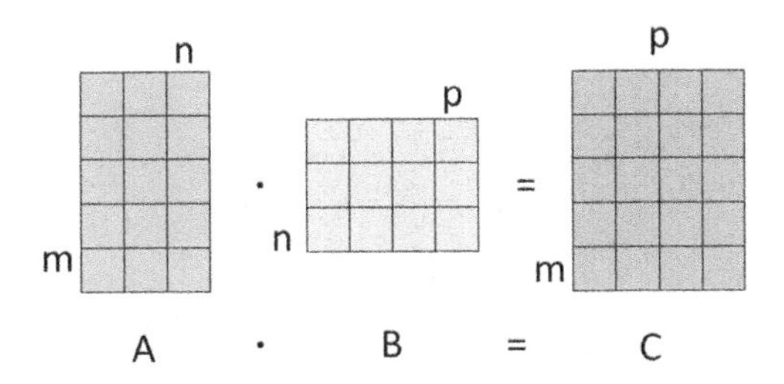

Figure 8.2 – The schematics of matrix multiplication

Let's look at an example. First, we have two vectors $|u\rangle$ and $|v\rangle$:

$$|u\rangle = \begin{bmatrix} 3 \\ 2 \\ -1 \end{bmatrix} \text{ and } |v\rangle = \begin{bmatrix} 7 \\ 2 \\ 3 \\ 1 \end{bmatrix}$$

Now, let's do the outer product with them:

$$|u\rangle\langle v| = \begin{bmatrix} 3 \\ 2 \\ -1 \end{bmatrix} \begin{bmatrix} 7 & 2 & 3 & 1 \end{bmatrix}$$

$$= \begin{bmatrix} 3\cdot 7 & 3\cdot 2 & 3\cdot 3 & 3\cdot 1 \\ 2\cdot 7 & 2\cdot 2 & 2\cdot 3 & 2\cdot 1 \\ -1\cdot 7 & -1\cdot 2 & -1\cdot 3 & -1\cdot 1 \end{bmatrix}$$

$$= \begin{bmatrix} 21 & 6 & 9 & 3 \\ 14 & 4 & 6 & 2 \\ -7 & -2 & -3 & -1 \end{bmatrix}$$

Is the outer product commutative? Well, let's try:

$$|v\rangle\langle u| = \begin{bmatrix} 7 \\ 2 \\ 3 \\ 1 \end{bmatrix} \begin{bmatrix} 3 & 2 & -1 \end{bmatrix}$$

$$= \begin{bmatrix} 7\cdot 3 & 7\cdot 2 & 7\cdot -1 \\ 2\cdot 3 & 2\cdot 2 & 2\cdot -1 \\ 3\cdot 3 & 3\cdot 2 & 3\cdot -1 \\ 1\cdot 3 & 1\cdot 2 & 1\cdot -1 \end{bmatrix}$$

$$= \begin{bmatrix} 21 & 14 & -7 \\ 6 & 4 & -2 \\ 9 & 6 & -3 \\ 3 & 2 & -1 \end{bmatrix}$$

Apparently not! So, in general, $|u\rangle\langle v| \neq |v\rangle\langle u|$. Look closely at the matrices for the final answers though. Do you see any similarities? That's right – they are the transpose of each other! This leads us to the following property of the outer product:

$$(|u\rangle\langle v|)^\dagger = (|v\rangle\langle u|)$$

You should also note that the outer product is distributive over addition, so:

$$\left(|v\rangle + |w\rangle \right)\langle u| = |v\rangle\langle u| + |w\rangle\langle u|$$

$$|u\rangle\left(\langle v| + \langle w| \right) = |u\rangle\langle v| + |u\rangle\langle w|. \tag{2}$$

So, let's try out *Equation (2)* on our friends $|0\rangle$ and $|1\rangle$. Here's the left side of the equation:

$$\left(|0\rangle + |1\rangle \right)\langle 1|$$

$$\left(\begin{bmatrix} 1 \\ 0 \end{bmatrix} + \begin{bmatrix} 0 \\ 1 \end{bmatrix} \right) \begin{bmatrix} 0 & 1 \end{bmatrix}$$

$$\begin{bmatrix} 1 \\ 1 \end{bmatrix} \begin{bmatrix} 0 & 1 \end{bmatrix}$$

$$\begin{bmatrix} 0 & 1 \\ 0 & 1 \end{bmatrix}$$

As an exercise, compute the right side of *Equation (2)*, and make sure that you get the same answer as you just got for the left side.

Exercise two

Compute the following:

$$|0\rangle\langle 1| + |1\rangle\langle 1|$$

Operators

In this section, we will consider linear operators. We first described these in *Chapter 5, Transforming Space with Matrices*. To reiterate, linear operators are linear transformations that map vectors from and to the *same vector space*. They are represented by square matrices. For just this section, I will put a "hat" or caret on the top of operators and use just the uppercase letter for matrices, as I want to be deliberate when referencing operators.

For instance, let's look at the $\hat{X}$ operator that transforms the zero and one states:

$$\hat{X}|0\rangle = |1\rangle$$
$$\hat{X}|1\rangle = |0\rangle$$

Now, let's come up with a matrix that represents this operator. The question becomes, which basis will we use? Let's use the computational basis, which is $|0\rangle$ and $|1\rangle$. I will denote this set of basis vectors by the letter C. So, the $\hat{X}$ operator in the C basis is represented by:

$$X_C = \begin{bmatrix} 0 & 1 \\ 1 & 0 \end{bmatrix}$$

Now, I want to come up with a matrix representation of $\hat{X}$ in the $|+\rangle$, $|-\rangle$ basis, which I will denote as +,-. Here is the matrix representation of the $\hat{X}$ operator in the +,- basis:

$$X_{+,-} = \begin{bmatrix} 1 & 0 \\ 0 & -1 \end{bmatrix}$$

Now that we have refreshed our knowledge around linear operators (which I will sometimes just call operators), let's look at another way we can represent an operator.

Representing an operator using the outer product

Can we use the outer product to represent an operator? Well, since the computation of an outer product results in a matrix, we most certainly can. But, given the previous discussion, since the outer product is a matrix, multiple outer products can represent an operator.

Let's show that we can by writing an equation in which a matrix A transforms a vector:

$$A|x\rangle = |y\rangle \tag{3}$$

Now, let's try to represent our matrix A by an outer product between two vectors $|u\rangle$ and $|v\rangle$:

$$A = |u\rangle\langle v|$$

Okay, let's rewrite *Equation (3)* using our new outer product representation of the matrix A:

$$\left(|u\rangle\langle v|\right)|x\rangle = |y\rangle$$

Due to the properties of the inner product and outer product that we have enumerated in the previous sections, we can rewrite this as:

$$|u\rangle\left(\langle v||x\rangle\right) = |y\rangle$$

Due to bra-ket notation, we can rewrite this with a bracket:

$$|u\rangle\left(\langle v|x\rangle\right) = |y\rangle \tag{4}$$

We know that a bracket denotes an inner product and therefore produces a scalar, like so:

$$\langle v|x\rangle = \alpha$$
$$\alpha \in \mathbb{C}$$

We can take our scalar α and put it back into *Equation (4)* and rearrange it:

$$|u\rangle\alpha = |y\rangle$$
$$\alpha|u\rangle = |y\rangle$$

We now see that the ket $|y\rangle$ that was a product of the matrix multiplication in *Equation (3)* is just another ket proportional to $|u\rangle$. This is great because it shows that we can represent a linear operator with an outer product in bra-ket notation. Not only that, the resulting ket will be proportional to the ket we use in the outer product.

So, how can this be used? Let's say we have an operator $\hat{O}$ defined with our plus and minus kets from before. This gives us a matrix O defined in an outer product, like so:

$$O = |+\rangle\langle-|$$

Now, we want to know what our new operator $\hat{O}$ does to the minus ket. We can do it without ever resorting to matrices, like so:

$$\hat{O}|-\rangle = O|-\rangle = (|+\rangle\langle-|)|-\rangle$$
$$\hat{O}|-\rangle = |+\rangle\langle-|-\rangle$$
$$\hat{O}|-\rangle = |+\rangle \cdot 1 = |+\rangle$$

So, our operator $\hat{O}$ turns the minus ket into the plus ket. You try it now.

Exercise 3

What does the operator $\hat{O}$ do to the plus ket? In other words, what is the following?

$$\hat{O}|+\rangle$$

The completeness relation

The following relation goes under a few names (the closure relation and the resolution of identity), but I will go with the name **completeness relation**. I will just state it without proving it, as the proof is rather laborious. We start with an orthonormal basis B where:

$$B = \left\{ |0\rangle, |1\rangle, |2\rangle, .., |n\rangle \right\},$$

Then, the identity operator $\hat{I}$ can be written as an outer product, like so:

$$I = \sum_{i=0}^{n} |i\rangle\langle i| \tag{5}$$

That's it. Doesn't seem like much, does it? But it is used over and over again in quantum computing to manipulate bra-ket expressions.

So, let's apply this to an example. Let's take our basis to be the plus and minus kets $\{|+\rangle, |-\rangle\}$. How then do we write the identity operator in this basis using the outer product representation? Well, using *Equation (5)*, we get:

$$I = |+\rangle\langle+| \ + \ |-\rangle\langle-|$$

Now, let's test the completeness relation. If we apply the identity operator to the plus ket, we should get the plus ket back. Let's see:

$$I|+\rangle = \left(|+\rangle\langle+| \; + \; |-\rangle\langle-| \right)|+\rangle$$

Distribute the plus ket

$$I|+\rangle = \left(|+\rangle\langle+|+\rangle \; + \; |-\rangle\langle-|+\rangle \right)$$

Calculate the inner products

$$I|+\rangle = \left(|+\rangle \cdot 1 \; + \; |-\rangle \cdot 0 \right) = |+\rangle$$

Indeed, we do get the plus ket back, and hence, we can see how the completeness relation works in a simple example.

The adjoint of an operator

As we saw, we can manipulate operators using bra-ket notation without resorting to a matrix representation. We also saw the use of the dagger symbol when looking at the conjugate transpose. Now, I'd like to take a step back and put this all together to define the **adjoint** of an operator.

In this definition, we need to remember that for every ket $|x\rangle$, there is a bra $\langle x|$. Let's say we have an operator $\hat{A}$ that acts on $|x\rangle$ to give us $|y\rangle$:

$$\hat{A}|x\rangle = |y\rangle,$$

Is there an associated operator we can use on the bra $\langle x|$ to give us the associated bra $\langle y|$of the ket $|y\rangle$? It happens that there is, and it is named the adjoint of the operator $\hat{A}$ and written this way:

$$\langle y| = \langle x|\hat{A}^\dagger$$

Now, there are rules when manipulating adjoint operators that you must keep in mind. First, the adjoint of a scalar is its complex conjugate:

$$\alpha^\dagger = \alpha^*$$
$$\alpha \in \mathbb{C}$$

You can distribute the adjoint operation amongst scalar multiplication of operators, like so:

$$(\alpha\hat{A})^\dagger = \alpha^*\hat{A}^\dagger$$

I will reiterate that bras and kets are the adjoints of each other:

$$(|x\rangle)^\dagger = \langle x|$$
$$(\langle x|)^\dagger = |x\rangle$$

If you take the adjoint of an adjoint, you get back the original operator:

$$(\hat{A}^\dagger)^\dagger = \hat{A}$$

You cannot distribute the adjoint across the multiplication of operators; rather, they anti-commute:

$$(\hat{A}\hat{B})^\dagger = \hat{B}^\dagger \hat{A}^\dagger$$

Finally, the adjoint can be distributed across a sum of operators:

$$(\hat{A} + \hat{B} + \hat{C})^\dagger = \hat{A}^\dagger + \hat{B}^\dagger + \hat{C}^\dagger$$

I know this seems like a lot of rules, but you will need them when you manipulate bra-ket expressions in quantum computing.

Finally, if an operator is expressed as an outer product, we can take its adjoint in the following way:

$$\hat{A} = |x\rangle\langle y|$$
$$\hat{A}^\dagger = |y\rangle\langle x|$$

When an operator is represented as a matrix *on an orthonormal basis*, then its adjoint is the conjugate transpose of the matrix:

$$\hat{A}^\dagger = A^\dagger$$

Let's look at a quick example. The $\hat{X}$ operator we have looked at previously can be represented in the computational basis as an outer product thusly:

$$X = |0\rangle\langle 1| + |1\rangle\langle 0|$$

Now, let's take the adjoint of X:

$$X^\dagger = (|0\rangle\langle 1| + |1\rangle\langle 0|)^\dagger$$
$$X^\dagger = (|0\rangle\langle 1|)^\dagger + (|1\rangle\langle 0|)^\dagger$$
$$X^\dagger = |1\rangle\langle 0| + |0\rangle\langle 1|$$

If you look closely, you should notice that the adjoint of X is equal to the original X. That's because $\hat{X}$ is Hermitian, which we will discuss in our next section.

Types of operators

There are certain types of linear operators that are special and need to be defined so that we can refer to them later in the book. You will also hear of them all the time in quantum computing.

Normal operators

Normal operators are ones that commute with their adjoint. For an operator $\hat{A}$, if:

$$\hat{A}\hat{A}^{\dagger} = \hat{A}^{\dagger}\hat{A} \tag{6}$$

then $\hat{A}$ is normal. They are important because a normal operator is diagonalizable, which is something we will consider later in the book. The following operators (Hermitian, unitary, positive, and positive semi-definite) are all normal operators.

A **normal matrix** represents a normal operator, and it commutes with its conjugate transpose. Let's look at an example normal matrix A:

$$A = \begin{bmatrix} -i & 2+3i \\ -2+3i & 0 \end{bmatrix}$$

Its conjugate transpose is:

$$A^{\dagger} = \begin{bmatrix} i & -2-3i \\ 2-3i & 0 \end{bmatrix}$$

Now, let's see if A commutes with its conjugate transpose. We'll calculate the left side of *Equation (6)* first:

$$AA^{\dagger} = \begin{bmatrix} -i & 2+3i \\ -2+3i & 0 \end{bmatrix} \begin{bmatrix} i & -2-3i \\ 2-3i & 0 \end{bmatrix}$$

$$AA^{\dagger} = \begin{bmatrix} (-i)i + (2+3i)(2-3i) & (-i)(-2-3i) + (2+3i)\cdot 0 \\ (-2+3i)i + 0\cdot(2-3i) & (-2+3i)(-2-3i) + 0\cdot 0 \end{bmatrix}$$

$$AA^{\dagger} = \begin{bmatrix} 14 & -3+2i \\ -3-2i & 13 \end{bmatrix}$$

And now we do the same for the right side of *Equation (6)*:

$$A^\dagger A = \begin{bmatrix} i & -2-3i \\ 2-3i & 0 \end{bmatrix} \begin{bmatrix} -i & 2+3i \\ -2+3i & 0 \end{bmatrix}$$

$$A^\dagger A = \begin{bmatrix} i(-i) + (-2-3i)(-2+3i) & i(2+3i) + (-2-3i)\cdot 0 \\ (2-3i)(-i) + 0\cdot(-2+3i) & (2-3i)(2+3i) + 0\cdot 0 \end{bmatrix}$$

$$A^\dagger A = \begin{bmatrix} 14 & -3+2i \\ -3-2i & 13 \end{bmatrix}$$

And there you go – the answers match and, therefore, our example matrix A is a normal matrix!

Normal operators and matrices have special properties, namely:

- They are **diagonalizable**; this will come up in the next chapter.
- Their eigenvalues are the conjugates of the eigenvalues of their adjoints.
- The eigenvectors associated with different eigenvalues are orthogonal.
- A vector space can be defined by an orthogonal basis set of these eigenvectors.

The operators that follow are all special types of normal operators.

Hermitian operators

The definition of a Hermitian operator is rather terse and simple as well. A Hermitian operator is an operator that is equal to its adjoint:

$$\hat{A} = \hat{A}^\dagger$$

You will also hear these referred to as self-adjoint operators.

Since Hermitian operators are normal, they have all their properties plus one more:

- All their eigenvalues are real numbers

Hermitian operators play an important part in quantum computing, as all measurements of a quantum state are done via a Hermitian operator.

Unitary operators

Unitary operators are very important, as they describe the evolution of a quantum state and therefore all gates in quantum computing are unitary. They also have a very simple definition:

$$\hat{U}^{-1} = \hat{U}^{\dagger}$$

Using this definition, we can also derive that:

$$\hat{U}^{-1}\hat{U} = \hat{U}\hat{U}^{-1} = \hat{I}$$
$$\hat{U}^{\dagger}\hat{U} = \hat{U}\hat{U}^{\dagger} = \hat{I}.$$

Unitary operators have two unique properties. First, their eigenvalues are complex numbers of modulus one:

$$|\lambda| = 1$$
$$\lambda = e^{i\theta}, \theta \in \mathbb{R}$$

And they preserve the inner product.

Let's quickly prove that they preserve the inner product using bra-ket notation:

$$\hat{U}|x\rangle = |x'\rangle, \hat{U}|y\rangle = |y'\rangle$$
$$\langle x'|y'\rangle = \left(\langle x|\hat{U}^{\dagger}\right)\left(\hat{U}|y\rangle\right) = \langle x|\underbrace{\hat{U}^{\dagger}\hat{U}}_{\hat{I}}|y\rangle$$
$$\langle x'|y'\rangle = \langle x|\hat{I}|y\rangle = \langle x|y\rangle$$

A consequence of unitary operators preserving the inner product is that they also preserve the norm of transformed vectors.

Unitary operators are represented by unitary matrices, and similarly:

$$U^{-1} = U^{\dagger}$$

In general, unitary matrices are expressed this way:

$$U = \begin{bmatrix} a & b \\ -e^{i\varphi}b^{*} & e^{i\varphi}a^{*} \end{bmatrix}, \quad |a|^2 + |b|^2 = 1$$

The determinant is:

$$\det(U) = e^{i\varphi}$$

Since all quantum computing gates are unitary, let's look at one and test it. The phase shift gate is represented in the computational basis by the following matrix:

$$P(\varphi) = \begin{bmatrix} 1 & 0 \\ 0 & e^{i\varphi} \end{bmatrix}$$

where φ is the phase shift with a period of 2π

To find the inverse of a matrix, we use our formula from *Chapter 7, Eigen-Stuff*:

$$A^{-1} = \begin{bmatrix} a & b \\ c & d \end{bmatrix}^{-1} = \frac{1}{\det(A)} \begin{bmatrix} d & -b \\ -c & a \end{bmatrix}$$

So, first, we have to find its determinant:

$$\det(P(\varphi)) = 1 \cdot e^{i\varphi} - 0 = e^{i\varphi}$$

Now, we just plug and play:

$$P^{-1}(\varphi) = \frac{1}{e^{i\varphi}} \begin{bmatrix} e^{i\varphi} & 0 \\ 0 & 1 \end{bmatrix} = \begin{bmatrix} 1 & 0 \\ 0 & \dfrac{1}{e^{i\varphi}} \end{bmatrix}$$

Now for the moment of truth – is it unitary?

$$P^{\dagger}(\varphi) = \begin{bmatrix} 1 & 0 \\ 0 & e^{-i\varphi} \end{bmatrix} = \begin{bmatrix} 1 & 0 \\ 0 & \dfrac{1}{e^{i\varphi}} \end{bmatrix}$$

Indeed, it is! Let's move on to projection operators.

Projection operators

We covered projection linear transformations in *Chapter 5, Using Matrices to Transform Space*. In that chapter, we defined a projection this way. If you have a linear transformation P, then if the following condition holds, it is a projection:

$$P^2 = P \tag{7}$$

In quantum computing, they are defined a little differently:

Given a normalized state $|\psi\rangle$the projection operator for this state is:

$$\hat{P} = |\psi\rangle\langle\psi| \tag{8}$$

Equation (7) still holds for projection operators, but given the definition in *Equation (8)*, you also get that the projection operator is Hermitian:

$$\hat{P} = \hat{P}^\dagger$$

All Projections Are Orthogonal in Quantum Computing

While there are non-orthogonal projection transformations in mathematics, we do not use them in quantum computing, so when you hear projection, assume that it is an orthogonal projection unless explicitly told otherwise.

If two projection operators commute, then their product is also a projection operator:

$$\text{If} \hat{P}_1\hat{P}_2 = \hat{P}_2\hat{P}_1 \text{then}$$
$$\hat{P}_1\hat{P}_2 \text{ is also a projection operator.}$$

In quantum computing, we often project one vector space onto another. Let's say I have a vector in an n-dimensional vector space defined by the basis $\{|0\rangle,|1\rangle,\ldots|n\rangle\}$ and I want to project it onto an m-dimensional subspace defined by the basis $\{|0\rangle,|1\rangle,\ldots|m\rangle\}$. Both bases are orthonormal. Then, the projection operator that projects onto our subspace is defined by:

$$P_m \equiv \sum_{i=1}^{m} |i\rangle\langle i|$$

The only eigenvalues a projection operator can have are zero and one. Now, let's move on to positive operators.

Positive operators

Positive operators are happy and optimistic, always looking on the bright side of life. Okay, maybe that's a joke :) Mathematically, positive operators are Hermitian. There are actually two types of positive operators, and they are only slightly different.

An operator $\hat{A}$ is said to be **positive definite** if:

$$\langle\psi|\hat{A}|\psi\rangle > 0 \tag{9}$$

Please consult the appendix on Bra-Ket notation if you are unfamiliar with the notation in *Equation (9)*.

An operator $\hat{A}$ is said to be **positive semidefinite** if:

$$\langle\psi|\hat{A}|\psi\rangle \geq 0$$

So the only difference is that positive definite operators do not include zero in their definition. All eigenvalues of positive operators are non-negative.

Okay, well, we're done with types of operators! There are quite a few, but they will come up in different segments of quantum computing, so try to keep them all straight.

> **No More Hats**
>
> As I said at the beginning of the section on operators, I will drop the hat or caret on top of operators, and you should be able to derive whether I mean an operator or a matrix from the context in which it is used.

Tensor products

Tensor products are a way to combine vector spaces. One of the postulates of quantum mechanics is that the state of a qubit is completely described by a unit vector in a Hilbert space. The problem then becomes how to deal with more than one qubit. This is where a tensor product comes in. Each qubit has its own Hilbert space, and to describe many qubits as a system, we need to combine all their Hilbert spaces into one bigger Hilbert space.

Mathematically, that means that if we have a Hilbert space H and another Hilbert space J, we denote their tensor product as:

$$M = H \otimes J$$

If H is an h dimensional space and J is a j dimensional space, then the dimension of the combined space M is $h \cdot j$. In other words:

$$\dim(M) = \dim(H) \cdot \dim(J)$$

Before we go any farther, let's look at the tensor product of two vectors.

The tensor product of vectors

The tensor product of two vectors is denoted in the following way in bra-ket notation. You'll notice that there are four different ways to notate it, so be careful when you come across it in quantum computing literature and know exactly what you are dealing with:

$$|u\rangle|v\rangle \quad or \quad |u\rangle \otimes |v\rangle \quad or \quad |uv\rangle \quad or \quad |u, v\rangle$$

Here are some of the properties of the tensor product:

- For a scalar s and a vector $|u\rangle$ in U and vector $|v\rangle$ in V, then:

$$s(|u\rangle \otimes |v\rangle) = (s|u\rangle) \otimes |v\rangle = |u\rangle \otimes (s|v\rangle)$$

- It is both right and left distributive. For vectors $|v\rangle$ and $|w\rangle$ in V and vectors $|u\rangle$ and $|z\rangle$ in U, then:

$$\left(|v\rangle + |w\rangle \right) \otimes |u\rangle = |v\rangle \otimes |u\rangle + |w\rangle \otimes |u\rangle$$
$$|v\rangle \otimes \left(|u\rangle + |z\rangle \right) = |v\rangle \otimes |u\rangle + |v\rangle \otimes |z\rangle$$

- It is not commutative, so in general:

$$|v\rangle \otimes |u\rangle \neq |u\rangle \otimes |v\rangle$$

Now, let's look at how we do the tensor product for actual column vectors. When the tensor product is implemented with arrays of numbers, it is actually the **Kronecker product**. Keep this in mind when researching and reading about the tensor product.

So, here is the Kronecker product defined mathematically for vectors. The tensor product of vectors is defined as:

$$\left|x\right\rangle \otimes \left|y\right\rangle = \left|xy\right\rangle = \begin{bmatrix} x_1 \\ x_2 \\ \vdots \\ x_h \end{bmatrix} \otimes \begin{bmatrix} y_1 \\ y_2 \\ \vdots \\ y_k \end{bmatrix} = \begin{bmatrix} x_1 \cdot y_1 \\ x_1 \cdot y_2 \\ \vdots \\ x_1 : y_k \\ x_2 \cdot y_1 \\ x_2 \cdot y_2 \\ \vdots \\ x_h \cdot y_1 \\ x_h \cdot y_2 \\ \vdots \\ x_h \cdot y_k \end{bmatrix}$$

In other words, take the first component of $|x\rangle$ and multiply it by every component of $|y\rangle$, then take the second component of $|x\rangle$ and multiply it by every component of $|y\rangle$, and repeat this procedure until all of the components of $|x\rangle$ are exhausted! That's a lot to say! Let's look at a definition with just *1* × *2* vectors:

$$\left|\psi\right\rangle = \begin{bmatrix} a \\ b \end{bmatrix} \otimes \begin{bmatrix} c \\ d \end{bmatrix} = \begin{bmatrix} ac \\ ad \\ bc \\ bd \end{bmatrix}$$

That's not bad, right? By the way, if you noticed, the tensor product of two vectors always produces another vector. Here is an example for you:

$$\begin{bmatrix} 2 \\ 0 \end{bmatrix} \otimes \begin{bmatrix} 3 \\ 1 \end{bmatrix} = \begin{bmatrix} 2 \cdot 3 \\ 2 \cdot 1 \\ 0 \cdot 3 \\ 0 \cdot 1 \end{bmatrix} = \begin{bmatrix} 6 \\ 2 \\ 0 \\ 0 \end{bmatrix}$$

Alright, how about some examples with our two favorite vectors $|0\rangle$ and $|1\rangle$? Let's do a tensor product between them:

$$|0\rangle \otimes |0\rangle = |00\rangle = \begin{bmatrix} 1 \\ 0 \end{bmatrix} \otimes \begin{bmatrix} 1 \\ 0 \end{bmatrix} = \begin{bmatrix} 1 \\ 0 \\ 0 \\ 0 \end{bmatrix} \qquad |0\rangle \otimes |1\rangle = |01\rangle = \begin{bmatrix} 1 \\ 0 \end{bmatrix} \otimes \begin{bmatrix} 0 \\ 1 \end{bmatrix} = \begin{bmatrix} 0 \\ 1 \\ 0 \\ 0 \end{bmatrix}$$

Exercise four

- What is $|11\rangle$?
- What is $|10\rangle$?

The basis of tensor product space

You will remember that we can completely describe a vector space by its basis. The same is true of a tensor product space. To get the basis of the tensor product space, you take the tensor product of each basis vector in one space with every basis vector in the other space involved in the tensor product. I hope you remember the Cartesian product from *Chapter 3, Foundations*. A quick reminder is that if I have two sets $A=\{x, y, z\}$ and $B=\{1, 2, 3\}$, their Cartesian product is all the ordered pairs shown in the following graphic:

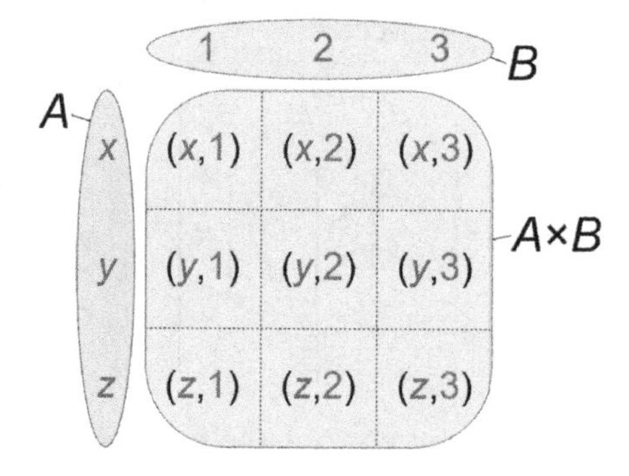

Figure 8.3 – An example of the Cartesian product of A × B [1]

So, another way to describe the basis of the new tensor product space is to first take the Cartesian product of all basis vectors and then do the tensor product between all the ordered pairs. And there you go!

Let's look at an example. What if we had two vector spaces, U and V, that were both two-dimensional? The basis for U is $\{|0\rangle,|1\rangle\}$, and the basis for V is $\{|+\rangle,|-\rangle\}$. Then, what is the basis for the tensor product of U and V? We have to calculate all the following tensor products:

$$|0\rangle \otimes |+\rangle$$
$$|0\rangle \otimes |-\rangle$$
$$|1\rangle \otimes |+\rangle$$
$$|1\rangle \otimes |-\rangle$$

So let's look at one of these and calculate it using the Kronecker product:

$$|0\rangle \otimes |+\rangle = \begin{bmatrix} 1 \\ 0 \end{bmatrix} \otimes \begin{bmatrix} \frac{1}{\sqrt{2}} \\ \frac{1}{\sqrt{2}} \end{bmatrix} = \begin{bmatrix} \frac{1}{\sqrt{2}} \\ \frac{1}{\sqrt{2}} \\ 0 \\ 0 \end{bmatrix}$$

So, that is one of the four vectors that are in the basis of our tensor product. Of course, once we have a basis, we can describe every vector in the space as a linear combination of our new basis vectors.

Exercise five

Calculate the Kronecker product of the following:

$$|0\rangle \otimes |-\rangle$$
$$|1\rangle \otimes |+\rangle$$
$$|1\rangle \otimes |-\rangle$$

The tensor product of operators

In bra-ket notation, there are rules to follow about the tensor product of linear operators. First, let's define the tensor product of operators in bra-ket notation. Let's say we have a vector $|v\rangle$ and linear operator A in vector space V. We also have a vector $|u\rangle$ and linear operator B in vector space U. Then, $A \otimes B$ on the tensor product space of $U \otimes V$ is defined as:

$$\left(A \otimes B \right)\left(|v\rangle \otimes |u\rangle \right) \equiv A|v\rangle \otimes B|u\rangle$$

Let's look at an example. We'll say we have two familiar linear operators X and Z and vector states $|0\rangle$ and $|1\rangle$. Here is the math of our example:

$$\begin{gathered} \left(X \otimes Z \right)\left(|0\rangle \otimes |1\rangle \right) = X|0\rangle \otimes Z|1\rangle \\ X|0\rangle = |1\rangle \text{ and } Z|1\rangle = -|1\rangle \text{ so} \\ X|0\rangle \otimes Z|1\rangle = |1\rangle \otimes -|1\rangle = -|11\rangle \end{gathered} \tag{10}$$

The Kronecker product of matrices

When we represent operators with matrices, we can calculate the tensor product using the Kronecker product. It is very similar to the Kronecker product for vectors, as column vectors are just n × 1 matrices. You basically take the first matrix entries and multiply them by the second matrix. Here's the definition. Given an m x n matrix A and a p × q matrix B, then their Kronecker product is an (m · p) × (n · q) matrix, like so:

$$\mathbf{A} \otimes \mathbf{B} = \begin{bmatrix} a_{11}\mathbf{B} & \cdots & a_{1n}\mathbf{B} \\ \vdots & \ddots & \vdots \\ a_{m1}\mathbf{B} & \cdots & a_{mn}\mathbf{B} \end{bmatrix}$$

That's pretty abstract; let's look at an example. Let's make matrix A a 2 × 2 matrix and B a 2 × 2 matrix as well. What will be the dimension of their Kronecker product? From the definition, we see it will be 4 × 4. Here's the example:

$$\begin{bmatrix} 1 & 1 \\ 3 & 4 \end{bmatrix} \otimes \begin{bmatrix} 0 & 5 \\ 6 & 0 \end{bmatrix} = \begin{bmatrix} 1\begin{bmatrix} 0 & 5 \\ 6 & 0 \end{bmatrix} & 1\begin{bmatrix} 0 & 5 \\ 6 & 0 \end{bmatrix} \\ 3\begin{bmatrix} 0 & 5 \\ 6 & 0 \end{bmatrix} & 4\begin{bmatrix} 0 & 5 \\ 6 & 0 \end{bmatrix} \end{bmatrix} = \begin{bmatrix} 1\cdot 0 & 1\cdot 5 & 1\cdot 0 & 1\cdot 5 \\ 1\cdot 6 & 1\cdot 0 & 1\cdot 6 & 1\cdot 0 \\ 3\cdot 0 & 3\cdot 5 & 4\cdot 0 & 4\cdot 5 \\ 3\cdot 6 & 3\cdot 0 & 4\cdot 6 & 4\cdot 0 \end{bmatrix} = \begin{bmatrix} 0 & 5 & 0 & 5 \\ 6 & 0 & 6 & 0 \\ 0 & 15 & 0 & 20 \\ 18 & 0 & 24 & 0 \end{bmatrix}$$

Exercise six

What is $B \otimes A$?

Now, let's redo our example from earlier, *Equation (10)*, in matrix form to double-check our work:

$$(X \otimes Z)(|0\rangle \otimes |1\rangle)$$

$$\left(\begin{bmatrix} 0 & 1 \\ 1 & 0 \end{bmatrix} \otimes \begin{bmatrix} 1 & 0 \\ 0 & -1 \end{bmatrix}\right)\left(\begin{bmatrix} 1 \\ 0 \end{bmatrix} \otimes \begin{bmatrix} 0 \\ 1 \end{bmatrix}\right)$$

$$\begin{bmatrix} 0 & 0 & 1 & 0 \\ 0 & 0 & 0 & -1 \\ 1 & 0 & 0 & 0 \\ 0 & -1 & 0 & 0 \end{bmatrix}\begin{bmatrix} 0 \\ 1 \\ 0 \\ 0 \end{bmatrix}$$

$$\begin{bmatrix} 0 \\ 0 \\ 0 \\ -1 \end{bmatrix} = -|1\rangle \otimes |1\rangle = -|11\rangle$$

Indeed, we get the same answer, which is assuring! Finally, let's look at the inner product for tensor product spaces.

The inner product of composite vectors

This is actually probably the most important definition when it comes to the tensor product because it enables us to have a Hilbert space as the output of our tensor product. Without further ado, here is the definition of the inner product for the tensor product of two vector spaces.

Two Hilbert spaces V and U create a tensor product space W:

$$W = V \otimes U$$

For vectors $|v\rangle$ and $|x\rangle$ in V and vectors $|u\rangle$ and $|z\rangle$ in U, we define two composite vectors in W:

$$|w\rangle = |v\rangle \otimes |x\rangle$$

$$|y\rangle = |u\rangle \otimes |z\rangle$$

We now define the inner product of these two composite vectors $|w\rangle$ and $|y\rangle$ as:

$$\langle w|y\rangle = \big(\langle v|\otimes\langle x|\big)\big(|u\rangle\otimes|z\rangle\big) = \langle v|u\rangle\langle x|z\rangle \tag{11}$$

We were able to simplify in *Equation (11)* due to the mixed-product property of the Kronecker product:

$$(A\otimes B)(C\otimes D) = (AC)\otimes(BD)$$

The inner product of two of our computation basis vectors in $\mathbb{C}^4$ should be zero. Let's check:

$$\langle 00|11\rangle = \big(\langle 0|\langle 0|\big)\big(|1\rangle|1\rangle\big) = \langle 0|1\rangle\langle 0|1\rangle = 0\cdot 0 = 0$$

Indeed, it is! Whew!

Exercise seven

Calculate the following inner products:

$$\langle 00|01\rangle$$
$$\langle 00|10\rangle$$
$$\langle 10|01\rangle$$

This concludes our discussion of tensor products.

Summary

We have covered a lot of ground in this chapter, but I hope you feel it's been worth it. I think it has because it brings everything we have worked on in the previous chapters together into one framework. Also, we can do real math with quantum computing now that we just couldn't do before. We will build on this in the last chapter to reach new heights in quantum computing that, at first, probably looked unattainable!

Answers to exercises

Exercise one

1. No, their inner product is -5.
2. Yes!

Exercise two

$$\begin{bmatrix} 0 & 1 \\ 0 & 1 \end{bmatrix}$$

Exercise three

$O|+\rangle = \mathbf{0}$

Exercise four

1. What is $|11\rangle$?

$$\begin{bmatrix} 0 \\ 0 \\ 0 \\ 1 \end{bmatrix}$$

2. What is $|10\rangle$?

$$\begin{bmatrix} 0 \\ 0 \\ 1 \\ 0 \end{bmatrix}$$

Exercise five

$$|0\rangle \otimes |-\rangle = \begin{bmatrix} \frac{1}{\sqrt{2}} \\ -\frac{1}{\sqrt{2}} \\ 0 \\ 0 \end{bmatrix}$$

$$|1\rangle \otimes |+\rangle = \begin{bmatrix} 0 \\ 0 \\ \frac{1}{\sqrt{2}} \\ \frac{1}{\sqrt{2}} \end{bmatrix}$$

$$|1\rangle \otimes |-\rangle = \begin{bmatrix} 0 \\ 0 \\ \frac{1}{\sqrt{2}} \\ -\frac{1}{\sqrt{2}} \end{bmatrix}$$

Exercise six

$$\begin{bmatrix} 0 & 0 & 5 & 5 \\ 0 & 0 & 15 & 20 \\ 6 & 6 & 0 & 0 \\ 18 & 24 & 0 & 0 \end{bmatrix}$$

Exercise seven

They all equal zero, as the states are orthogonal to each other.

9
Advanced Concepts

In this chapter, we will go into some advanced linear algebra concepts. These will not come up all the time in quantum computing, but when they do, you should know what they are and where to find information about them. Almost all the topics are about decomposing a matrix. This becomes important in quantum computing because when we come up with a unitary transformation that we'd like to do on a quantum computer, we will only have certain unitary operators to use on it. Then, it becomes a question of which combination of available operators we should use so that we can perform our overall unitary transformation. Along the way, we will also look at important inequalities and how to represent functions that have matrices in them.

In this chapter, we are going to cover the following main topics:

- Gram-Schmidt
- Cauchy-Schwarz and triangle inequalities
- Spectral decomposition
- Singular value decomposition
- Polar decomposition
- Operator functions and the matrix exponential

Gram-Schmidt

The **Gram-Schmidt process** is an algorithm in which you input a basis set of vectors and it outputs a basis set that is orthogonal. We can then normalize that set of vectors, and suddenly, we have an orthonormal set of basis vectors! This is very helpful in quantum computing and other areas of applied math, as an orthonormal basis is usually the best basis for computations and representing vectors with coordinates.

> **Gram-Schmidt Is a Decomposition Tool**
>
> While we won't go into it in this book, the Gram-Schmidt process is used in certain decompositions, so it's good to know from that vantage point too.

Let's look at an example before getting into the nitty-gritty of the actual procedure (which can be dry and dull). Let's say I have a basis for $\mathbb{C}^2$, such as the following:

$$|x_1\rangle = \begin{bmatrix} 2 \\ 0 \end{bmatrix} \quad |x_2\rangle = \begin{bmatrix} 1 \\ -2i \end{bmatrix}$$

These vectors are not orthogonal, since their inner product does not equal 0:

$$\langle x_1 | x_2 \rangle = \begin{bmatrix} 2 & 0 \end{bmatrix} \begin{bmatrix} 1 \\ -2i \end{bmatrix} = 2 - 0 = 2$$

They are also not normalized. Now, I want to get an orthonormal basis for $\mathbb{C}^2$ using these two vectors. Here's the process. The first step is the easiest; we just choose the first vector in our set, $|x_1\rangle$, to be the first vector in our soon to be orthogonal basis set (denoted by $|v_1\rangle$):

$$|v_1\rangle = |x_1\rangle = \begin{bmatrix} 2 \\ 0 \end{bmatrix}$$

That was easy enough. This is the easiest step of Gram-Schmidt, which is always the first step. Now for the second step. Here is the formula:

$$|v_2\rangle = |x_2\rangle - \frac{\langle x_2 | v_1 \rangle}{\langle v_1 | v_1 \rangle} |v_1\rangle$$

Let's calculate the numerator of the right part of the equation first:

$$\langle x_2 | v_1 \rangle = \begin{bmatrix} 1 & 2i \end{bmatrix} \begin{bmatrix} 2 \\ 0 \end{bmatrix} = 2$$

Now for the denominator:

$$\langle v_1 | v_1 \rangle = \begin{bmatrix} 2 & 0 \end{bmatrix} \begin{bmatrix} 2 \\ 0 \end{bmatrix} = 4$$

Finally, let's put it all together!

$$|v_2\rangle = |x_2\rangle - \frac{\langle x_2 | v_1 \rangle}{\langle v_1 | v_1 \rangle} |v_1\rangle$$

$$\frac{\langle x_2 | v_1 \rangle}{\langle v_1 | v_1 \rangle} = \frac{2}{4} = \frac{1}{2}$$

$$|v_2\rangle = \begin{bmatrix} 1 \\ -2i \end{bmatrix} - \frac{1}{2}\begin{bmatrix} 2 \\ 0 \end{bmatrix}$$

$$|v_2\rangle = \begin{bmatrix} 1 \\ -2i \end{bmatrix} - \begin{bmatrix} 1 \\ 0 \end{bmatrix} = \begin{bmatrix} 0 \\ -2i \end{bmatrix}$$

So, our new orthogonal basis for $\mathbb{C}^2$ is the following basis set:

$$|v_1\rangle = \begin{bmatrix} 2 \\ 0 \end{bmatrix}, \ |v_2\rangle = \begin{bmatrix} 0 \\ -2i \end{bmatrix}$$

You should calculate their inner product to ensure it is zero and that they are indeed orthogonal. So, now all we need to do is normalize them to get an orthonormal basis! Let's do $|v_1\rangle$ first:

$$\| v_1 \|^2 = \langle v_1 | v_1 \rangle = 4$$

$$\| v_1 \| = \sqrt{4} = 2$$

$$|e_1\rangle = \frac{|v_1\rangle}{\| v_1 \|} = \begin{bmatrix} 2 \\ 0 \end{bmatrix} \cdot \frac{1}{2} = \begin{bmatrix} 1 \\ 0 \end{bmatrix}$$

Now for $|v_2\rangle$:

$$\| v_2 \|^2 = \langle v_2 | v_2 \rangle = \begin{bmatrix} 0 & 2i \end{bmatrix} \begin{bmatrix} 0 \\ -2i \end{bmatrix} = 4$$

$$\| v_2 \| = \sqrt{4} = 2$$

$$|e_2\rangle = \frac{|v_2\rangle}{\| v_2 \|} = \begin{bmatrix} 0 \\ -2i \end{bmatrix} \cdot \frac{1}{2} = \begin{bmatrix} 0 \\ -i \end{bmatrix}$$

We now have an orthonormal basis for $\mathbb{C}^2$, as shown in the following:

$$|e_1\rangle = \begin{bmatrix} 1 \\ 0 \end{bmatrix} \quad , \quad |e_2\rangle = \begin{bmatrix} 0 \\ -i \end{bmatrix}$$

And this all started from two linearly independent vectors using the Gram-Schmidt process. Pretty cool, eh?

Now, it's time to lay out the entire process:

Given a basis $B = \{ |x_1\rangle, |x_2\rangle, \ldots, |x_p\rangle \}$ for a vector space V, define

$$|v_1\rangle = |x_1\rangle$$

$$|v_2\rangle = |x_2\rangle - \frac{\langle x_2 | v_1 \rangle}{\langle v_1 | v_1 \rangle} |v_1\rangle$$

$$|v_3\rangle = |x_3\rangle - \frac{\langle x_3 | v_1 \rangle}{\langle v_1 | v_1 \rangle} |v_1\rangle - \frac{\langle x_3 | v_2 \rangle}{\langle v_2 | v_2 \rangle} |v_2\rangle$$

$$\vdots$$

$$|v_p\rangle = |x_p\rangle - \frac{\langle x_p | v_1 \rangle}{\langle v_1 | v_1 \rangle} |v_1\rangle - \frac{\langle x_p | v_2 \rangle}{\langle v_2 | v_2 \rangle} |v_2\rangle - \ldots - \frac{\langle x_p | v_{p-1} \rangle}{\langle v_{p-1} | v_{p-1} \rangle}$$

Then $\{ |v_1\rangle, |v_2\rangle, \ldots, |v_p\rangle \}$ is an orthogonal basis for V.

This algorithm will give us an orthogonal basis; then, all we have to do is normalize each vector, and we get an orthonormal basis! Let's move on to two important inequalities.

Cauchy-Schwarz and triangle inequalities

The **Cauchy-Schwarz inequality** is one of the most important inequalities in mathematics. Succinctly stated, it says that the absolute value of the inner product of two vectors is less than or equal to the norm of those two vectors multiplied together. In fact, they are only equal if the two vectors are linearly dependent:

$$\left|\langle u|v\rangle\right| \leq \left\|u\right\|\left\|v\right\|$$

There are several proofs of this inequality, which I encourage you to seek out if you are interested. But, in the totality of things, knowing this inequality is all that is really required for quantum computing.

The other major inequality is the **triangle inequality**. It comes from our old friend Euclid in his book *The Elements*. Succinctly stated, it says that the length of two sides of a triangle must always be more than the length of one side. They will only be equal in the corner case when the triangle has zero area. It is very intuitive once you see some example triangles. Here are some triangles that show how the side z is less than the sum of sides x and y:

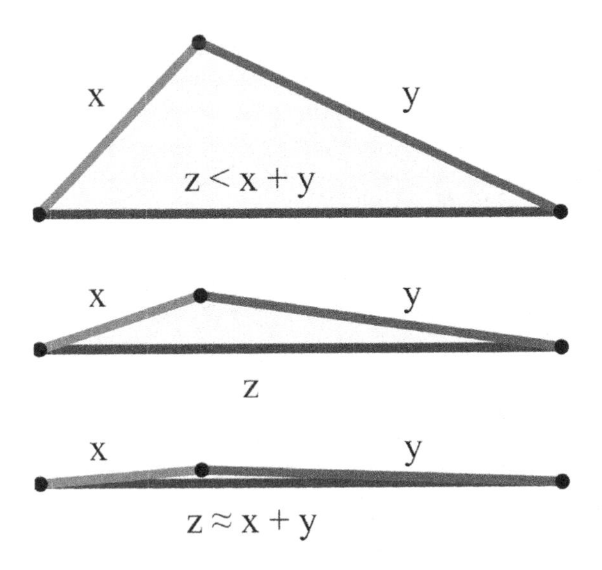

Figure 9.1 – Three example triangles for the triangle inequality [1]

Now, let's state the triangle inequality:

$$\| x+y \| \leq \| x \| + \| y \|$$

You'll notice that the way it is notated is using the norms of vectors. Let's look at an example. Let's say we have the two following kets:

$$|x\rangle = 2|0\rangle + 3i|1\rangle$$
$$|y\rangle = 2i|0\rangle - |1\rangle$$

Their corresponding bras (that is, their conjugate transpose) are:

$$\langle x| = 2\langle 0| - 3i\langle 1|$$
$$\langle y| = -2i\langle 0| - \langle 1|$$

Now, the norm of $|x\rangle$ is:

$$\| x \|^2 = \langle x|x\rangle = \left(2\langle 0| - 3i\langle 1| \right)\left(2|0\rangle + 3i|1\rangle \right)$$
$$\| x \|^2 = \langle x|x\rangle = 4\langle 0|0\rangle + 6i\cancelto{0}{\langle 0|1\rangle} - 6i\cancelto{0}{\langle 1|0\rangle} + 9\langle 1|1\rangle$$
$$\| x \|^2 = \langle x|x\rangle = 4 + 9 = 13$$
$$\| x \| = \sqrt{13}$$

The norm of $|y\rangle$ is:

$$\| y \|^2 = \langle y|y\rangle = \left(-2i\langle 0| - \langle 1| \right)\left(2i|0\rangle - |1\rangle \right)$$
$$\| y \|^2 = \langle y|y\rangle = 4\langle 0|0\rangle + 2i\cancelto{0}{\langle 0|1\rangle} - 2i\cancelto{0}{\langle 1|0\rangle} + 1\langle 1|1\rangle$$
$$\| y \|^2 = \langle y|y\rangle = 4 + 1 = 5$$
$$\| y \| = \sqrt{5}$$

So, the right side of the triangle inequality is:

$$\| x \| + \| y \| = \sqrt{13} + \sqrt{5} \approx 5.84$$

Let's calculate the left side of the triangle inequality:

$$|x\rangle + |y\rangle = 2|0\rangle + 3i|1\rangle + 2i|0\rangle - |1\rangle$$
$$|x\rangle + |y\rangle = (2+2i)|0\rangle + (-1+3i)|1\rangle$$

$$\|x+y\|^2 = \big((2-2i)\langle 0| + (-1-3i)\langle 1| \big)\big((2+2i)|0\rangle + (-1+3i)|1\rangle \big)$$
$$\|x+y\|^2 = (4+4)\langle 0|0\rangle + (1+9)\langle 1|1\rangle$$
$$\|x+y\|^2 = 8+10 = 18$$
$$\|x+y\| = \sqrt{18} = 3\sqrt{2} \approx 4.24$$

So, the triangle equality holds for our two kets $|x\rangle$ and $|y\rangle$ since $4.24 \le 5.84$. Now, we will move to look at decompositions of matrices.

Spectral decomposition

The **spectrum** of a square matrix is the set of its eigenvalues. There is a cool theorem in linear algebra that states that all matrices representing a linear operator have the *same* spectrum. Before we use the spectrum though, we need to talk about diagonal matrices.

Diagonal matrices

The **main diagonal** of a matrix is every entry where the row index equals the column index. Examples make this very easy to see. All the following matrices have the letter *d* on their main diagonal:

$$\begin{bmatrix} d & 3 \\ 2 & d \end{bmatrix} \quad \begin{bmatrix} d & 2 & 2 \\ 4 & d & 4 \\ 2 & 1 & d \end{bmatrix} \quad \begin{bmatrix} d & 4 & 2 & 3 \\ 4 & d & 2 & 4 \\ 5 & 1 & d & 4 \\ 1 & 2 & 2 & d \end{bmatrix}$$

Now, a **diagonal matrix** has zero on all entries outside the main diagonal. Here are examples of diagonal matrices:

$$\begin{bmatrix} 2 & 0 \\ 0 & i \end{bmatrix} \quad \begin{bmatrix} 3 & 0 & 0 \\ 0 & -2 & 0 \\ 0 & 0 & 4 \end{bmatrix} \quad \begin{bmatrix} 3 & 0 & 0 & 0 \\ 0 & 3 & 0 & 0 \\ 0 & 0 & i & 0 \\ 0 & 0 & 0 & 5 \end{bmatrix}$$

Here are two cool features of diagonal matrices that make them all the rage at linear algebra parties. One, all their eigenvalues are on their main diagonal. Two, they are very easy to exponentiate. Let's see the latter in action real quick:

$$\begin{bmatrix} 2 & 0 \\ 0 & 2 \end{bmatrix}^2 = \begin{bmatrix} 2^2 & 0 \\ 0 & 2^2 \end{bmatrix} = \begin{bmatrix} 4 & 0 \\ 0 & 4 \end{bmatrix}$$

$$\begin{bmatrix} 2 & 0 \\ 0 & 2 \end{bmatrix}^3 = \begin{bmatrix} 2^3 & 0 \\ 0 & 2^3 \end{bmatrix} = \begin{bmatrix} 8 & 0 \\ 0 & 8 \end{bmatrix}$$

$$\vdots$$

$$\begin{bmatrix} 2 & 0 \\ 0 & 2 \end{bmatrix}^n = \begin{bmatrix} 2^n & 0 \\ 0 & 2^n \end{bmatrix}$$

Try to exponentiate any regular old random matrix... it'll take you a while! Now, back to the main feature – spectral decomposition.

Spectral theory

The **spectral theorem** lets us know when an operator can be represented by a diagonal matrix. These operators and all the matrices that represent them are called **diagonalizable**. This means that we can factor a diagonalizable matrix A into:

$$A = PDP^{-1}$$

where P is an invertible matrix and D is a diagonal matrix.

If you remember from the last chapter, all normal operators and their associated normal matrices are diagonalizable. If this is the case, then we can decompose a matrix A into even more special factors:

$$A = U\Lambda U^\dagger$$

Looks like a fraternity name, right? Well, U is a unitary matrix, and the capital Greek letter lambda (Λ) is a diagonal matrix with all the eigenvalues of A on its main diagonal! Remember that we used the lowercase lambda (λ) for each eigenvalue, so it makes sense to use the uppercase lambda for all of them together in one matrix. Now, here comes the kicker – the column vectors making up U are the eigenvectors of A!

Putting this all together means that we can decompose any normal matrix into a unitary matrix U, its conjugate transpose, and a diagonal matrix Λ that has the eigenvalues down its main diagonal. Please note that the eigenvalues placement on the main diagonal must correspond with its eigenvector's placement in U. Since the set of eigenvalues is called the spectrum, I hope you can see why this is called spectral decomposition.

Okay, enough talk – more action! Let's do an example.

> **Diagonalizable Matrices**
> Please note that there are diagonalizable matrices that are not normal, but we do not often come across these matrices in quantum computing.

Example

Let's use spectral decomposition to decompose the quantum gate Y. Here is its matrix in the computational basis:

$$\begin{bmatrix} 0 & -i \\ i & 0 \end{bmatrix}$$

Now, I will go ahead and tell you that the eigenvalues of Y are 1 and -1. The eigenvectors are as follows where "+" denotes the eigenvector for the eigenvalue of 1 and "–" denotes the eigenvector of -1:

$$|\lambda_{Y+}\rangle = \begin{bmatrix} \frac{1}{\sqrt{2}} \\ \frac{i}{\sqrt{2}} \end{bmatrix} \quad , \quad |\lambda_{Y-}\rangle = \begin{bmatrix} \frac{1}{\sqrt{2}} \\ \frac{-i}{\sqrt{2}} \end{bmatrix}$$

So, according to the spectral theorem, I can represent Y in terms of a matrix U that has its eigenvectors as the columns, a diagonal matrix with the eigenvalues down its main diagonal, and the conjugate transpose of U. So, here is U and U dagger:

$$U = \begin{bmatrix} \frac{1}{\sqrt{2}} & \frac{1}{\sqrt{2}} \\ \frac{i}{\sqrt{2}} & \frac{-i}{\sqrt{2}} \end{bmatrix} \quad , \quad U^{\dagger} = \begin{bmatrix} \frac{1}{\sqrt{2}} & \frac{-i}{\sqrt{2}} \\ \frac{1}{\sqrt{2}} & \frac{i}{\sqrt{2}} \end{bmatrix}$$

Here is the eigenvalue matrix:

$$\Lambda = \begin{bmatrix} 1 & 0 \\ 0 & -1 \end{bmatrix}$$

Putting this all together, we can now decompose Y into:

$$Y = U\Lambda U^{\dagger} = \begin{bmatrix} \frac{1}{\sqrt{2}} & \frac{1}{\sqrt{2}} \\ \frac{i}{\sqrt{2}} & \frac{-i}{\sqrt{2}} \end{bmatrix} \begin{bmatrix} 1 & 0 \\ 0 & -1 \end{bmatrix} \begin{bmatrix} \frac{1}{\sqrt{2}} & \frac{-i}{\sqrt{2}} \\ \frac{1}{\sqrt{2}} & \frac{i}{\sqrt{2}} \end{bmatrix}$$

I encourage you to work out the matrix multiplication and make sure I'm right!

But, hey – wait a minute. We're in quantum computing and we like bra-ket notation! So, let's do it in bra-ket notation!

Bra-ket notation

All the way back in *Chapter 2, The Matrix*, we talked about how you can represent a matrix as a set of column vectors (kets) or row vectors (bras). Well, we're going to use that here. I'll prove this using 2 × 2 matrices, and then you'll have to trust me that it works for n × n matrices. I can represent the unitary matrix U that has eigenvectors for its column vectors like so:

$$U = \begin{bmatrix} |\lambda_1\rangle & |\lambda_2\rangle \\ & \end{bmatrix}$$

U dagger then becomes:

$$U^{\dagger} = \begin{bmatrix} \langle\lambda_1| & \\ \langle\lambda_2| & \end{bmatrix}$$

The eigenvalue matrix will look like this:

$$\Lambda = \begin{bmatrix} \lambda_1 & 0 \\ 0 & \lambda_2 \end{bmatrix}$$

Putting it all together, it looks like:

$$A = U\Lambda U^{\dagger} = \begin{bmatrix} |\lambda_1\rangle & |\lambda_2\rangle \end{bmatrix} \begin{bmatrix} \lambda_1 & 0 \\ 0 & \lambda_2 \end{bmatrix} \begin{bmatrix} \langle\lambda_1| \\ \langle\lambda_2| \end{bmatrix}$$

Multiplying the first two matrices, we get:

$$A = U\Lambda U^{\dagger} = \begin{bmatrix} \lambda_1|\lambda_1\rangle & \lambda_2|\lambda_2\rangle \end{bmatrix} \begin{bmatrix} \langle\lambda_1| \\ \langle\lambda_2| \end{bmatrix}$$

Then, when we do the final multiplication, we get:

$$A = U\Lambda U^{\dagger} = \lambda_1|\lambda_1\rangle\langle\lambda_1| + \lambda_2|\lambda_2\rangle\langle\lambda_2|$$

And this is the bra-ket equation for spectral decomposition! For an n × n matrix, it becomes:

$$A = \sum_{i=1}^{n} \lambda_i |\lambda_i\rangle\langle\lambda_i|$$

This is a very important result! It means that any normal operator can be represented as a linear combination of outer products composed of just its eigenvalues and eigenvectors! Let's see how this plays out with our *Y* operator.

Example take two

You may have noticed that the eigenvectors of *Y* look eerily familiar; that's because they are the *i* and minus *i* states!

$$|\lambda_{Y+}\rangle = |i\rangle \equiv \begin{bmatrix} \frac{1}{\sqrt{2}} \\ \frac{i}{\sqrt{2}} \end{bmatrix} \quad , \quad |\lambda_{Y-}\rangle = |-i\rangle \equiv \begin{bmatrix} \frac{1}{\sqrt{2}} \\ \frac{-i}{\sqrt{2}} \end{bmatrix}$$

So, using spectral decomposition, we can represent the quantum *Y* gate this way:

$$Y = |i\rangle\langle i| - |-i\rangle\langle -i| \qquad (1)$$

If you want to express this in the computational basis, you have to write *i* and minus *i* in it:

$$|i\rangle = \frac{1}{\sqrt{2}}\left(|0\rangle + i|1\rangle \right)$$

$$|-i\rangle = \frac{1}{\sqrt{2}}\left(|0\rangle - i|1\rangle \right)$$

Then, substitute this with *Equation (1)* and work out the math. You will get:

$$Y = i\left(|1\rangle\langle 0| - |0\rangle\langle 1| \right)$$

Now, let's look at another decomposition.

Singular value decomposition

Singular Value Decomposition (**SVD**) is probably the most famous decomposition you can do for linear operators and matrices. It is at the core of search engines and machine learning algorithms. Additionally, it can be used on any type of matrix, even rectangular ones. However, we will only look at square matrices.

Succinctly stated, it guarantees that for any matrix A, it can be decomposed into three matrices:

$$A = U\Sigma V^{\dagger},$$

Whereas U is a unitary matrix, Σ (sigma) is a diagonal matrix with what is known as the **singular values** of A on its diagonal, and V is also a unitary matrix. It should be noted that this decomposition is not unique, and different matrices can be used for U, Σ, and V.

Let's look at an example. We have the following matrix A:

$$A = \begin{bmatrix} -2 & 0 \\ 0 & 0 \end{bmatrix}$$

Without going through the math, I'm going to tell you that SVD can be used to get this decomposition:

$$A = \begin{bmatrix} -1 & 0 \\ 0 & 1 \end{bmatrix}\begin{bmatrix} 2 & 0 \\ 0 & 0 \end{bmatrix}\begin{bmatrix} 1 & 0 \\ 0 & 1 \end{bmatrix}$$

Let's make sure that U and V are unitary matrices:

$$U = \begin{bmatrix} -1 & 0 \\ 0 & 1 \end{bmatrix} \quad , \quad U^{\dagger} = \begin{bmatrix} -1 & 0 \\ 0 & 1 \end{bmatrix}$$

$$\text{Verify that } UU^{\dagger} = I$$

$$\begin{bmatrix} -1 & 0 \\ 0 & 1 \end{bmatrix}\begin{bmatrix} -1 & 0 \\ 0 & 1 \end{bmatrix} = \begin{bmatrix} 1 & 0 \\ 0 & 1 \end{bmatrix} = I$$

Since V is the identity matrix, I think we can safely assume it is unitary.

Now for these pesky singular values on the main diagonal of Σ. How do we get those? Those are found by taking the square root of the eigenvalues of A multiplied by its conjugate transpose. So for us, this is:

$$A = \begin{bmatrix} -2 & 0 \\ 0 & 0 \end{bmatrix} \quad , \quad A^{\dagger} = \begin{bmatrix} -2 & 0 \\ 0 & 0 \end{bmatrix}$$

$$AA^{\dagger} = \begin{bmatrix} -2 & 0 \\ 0 & 0 \end{bmatrix}\begin{bmatrix} -2 & 0 \\ 0 & 0 \end{bmatrix} = \begin{bmatrix} 4 & 0 \\ 0 & 0 \end{bmatrix}$$

The eigenvalues are on the main diagonal, and they are 4 and 0. So, the singular values of A are the square root of these eigenvalues, namely 2 and 0. If you look at the middle matrix, Σ, you'll notice that it has these singular values on its main diagonal. Your next question might be, how do we find U and V? Well, you may be disappointed, but I'm not going to go through the algorithm here. Suffice it to say that it involves finding an orthonormal set of eigenvectors for $AA^{\dagger}$. If you are interested in learning more, I encourage you to look into one of the linear algebra books in my appendix of references. Though, to be quite truthful, we almost always use computers to calculate SVD. Let's move onto another decomposition – polar decomposition.

Polar decomposition

Polar decomposition allows you to factor any matrix into unitary and positive semi-definite Hermitian matrices. It can be seen as breaking down a linear transformation into a rotation or reflection and scaling in $\mathbb{R}^n$. Formally, it is as follows:

$$A = UP,$$

for any matrix A. U is a unitary matrix and P is a positive semi-definite matrix. Let's look at an example:

$$A = \begin{bmatrix} 2 & 0 \\ 0 & -1 \end{bmatrix}$$

Using polar decomposition, this matrix can be decomposed into:

$$A = UP = \begin{bmatrix} 1 & 0 \\ 0 & -1 \end{bmatrix} \begin{bmatrix} 2 & 0 \\ 0 & 1 \end{bmatrix}$$

This may not seem like much, but we took a random matrix and turned it into a reflection matrix times a scaling matrix. Pretty cool!

Again, I will not go through the algorithm here because we will use calculators. Calculators for polar decomposition are not as plentiful as SVD, but I have found using the SciPy Python library to be the best way.

Operator functions and the matrix exponential

Now, we get to look at functions that involve an operator and their matrix representations. What types of functions are we talking about? Well, really any function that can be defined, such as the sine of x. You have probably never seen a function like this:

$$f(A) = \sin(A)$$

where A is a matrix. The first question is, does this even make sense? Well, mathematicians have come up with ways for this to make sense, and it has applications in quantum computing.

As we have said, if a matrix A is diagonalizable, it can be decomposed into an invertible matrix P and diagonal matrix D as:

$$A = PDP^{-1} \tag{2}$$

Given that, we can represent a function involving such a matrix like so:

$$f(A) = P \begin{bmatrix} f(d_1) & \cdots & 0 \\ \vdots & \ddots & \vdots \\ 0 & \cdots & f(d_n) \end{bmatrix} P^{-1} \quad (3)$$

where the matrix in the middle is the diagonal matrix D in *Equation (2)*. The function is evaluated for every value on the main diagonal of D.

Let's look at an example. Let's do the easiest case, where we are already dealing with a diagonal matrix:

$$A = \begin{bmatrix} \pi & 0 \\ 0 & \pi/2 \end{bmatrix}$$

Now, we want to find the sine of A. Following the process from *Equation (3)*, this is what we get:

$$\sin(A) = \begin{bmatrix} \sin \pi & 0 \\ 0 & \sin \pi/2 \end{bmatrix}$$

$$\sin(A) = \begin{bmatrix} 0 & 0 \\ 0 & 1 \end{bmatrix}$$

How about we check our answer by finding the cosine of A as well and then summing their squares? Here is the cosine of A:

$$\cos(A) = \begin{bmatrix} \cos \pi & 0 \\ 0 & \cos \pi/2 \end{bmatrix}$$

$$\cos(A) = \begin{bmatrix} -1 & 0 \\ 0 & 0 \end{bmatrix}$$

We know the trigonometric identity:

$$\sin^2 x + \cos^2 x = 1$$

The number 1 in that identity becomes the identity matrix for us when dealing with matrices. Now, let's verify our calculations for sine and cosine!

$$\sin(A) = \begin{bmatrix} 0 & 0 \\ 0 & 1 \end{bmatrix}$$

$$\sin^2(A) = \begin{bmatrix} 0 & 0 \\ 0 & 1 \end{bmatrix}$$

$$\cos(A) = \begin{bmatrix} -1 & 0 \\ 0 & 0 \end{bmatrix}$$

$$\cos^2(A) = \begin{bmatrix} 1 & 0 \\ 0 & 0 \end{bmatrix}$$

$$\sin^2(A) + \cos^2(A) = \begin{bmatrix} 0 & 0 \\ 0 & 1 \end{bmatrix} + \begin{bmatrix} 1 & 0 \\ 0 & 0 \end{bmatrix} = \begin{bmatrix} 1 & 0 \\ 0 & 1 \end{bmatrix} = I$$

So, our calculations are correct and this makes sense!

A function that comes up a lot in quantum computing is the exponential function:

$$f(x) = e^x = \exp(x)$$

You'll notice that we use $\exp(x)$ to also denote the exponential function, as it is easier to write sometimes. If the matrix is diagonal, such as the Z operator, this becomes easy to calculate:

$$Z = \begin{bmatrix} 1 & 0 \\ 0 & -1 \end{bmatrix}$$

$$e^Z = \begin{bmatrix} e^1 & 0 \\ 0 & e^{-1} \end{bmatrix}$$

Let's see a slightly harder example. What is $\exp(A)$ when A is:

$$A = \begin{bmatrix} 1 & 2 \\ 0 & -1 \end{bmatrix}$$

Well, first we need to diagonalize it, and the first step of that is finding its eigenvalues:

$$\begin{bmatrix} 1 & 2 \\ 0 & -1 \end{bmatrix} - \lambda \begin{bmatrix} 1 & 0 \\ 0 & 1 \end{bmatrix} = \begin{bmatrix} 1-\lambda & 2 \\ 0 & -1-\lambda \end{bmatrix}$$

$$\det \begin{bmatrix} 1-\lambda & 2 \\ 0 & -1-\lambda \end{bmatrix} = (1-\lambda)(-1-\lambda) - 2 \cdot 0$$

$$\det \begin{bmatrix} 1-\lambda & 2 \\ 0 & -1-\lambda \end{bmatrix} = \lambda^2 - 1$$

$$\lambda^2 = 1$$

$$\lambda_1 = 1 \text{ , } \lambda_2 = -1$$

Then, we have to find its eigenvectors. I will go ahead and tell you that they are:

$$|\lambda_1\rangle = \begin{bmatrix} 1 \\ 0 \end{bmatrix} \text{ , } |\lambda_2\rangle = \begin{bmatrix} -1 \\ 1 \end{bmatrix}$$

Given all of this, we can write its diagonal representation like so:

$$A = PDP^{-1} = \begin{bmatrix} 1 & -1 \\ 0 & 1 \end{bmatrix} \begin{bmatrix} 1 & 0 \\ 0 & -1 \end{bmatrix} \begin{bmatrix} 1 & 1 \\ 0 & 1 \end{bmatrix}$$

Evaluating the exponential function on the main diagonal of D gives us:

$$\exp(A) = \exp(PDP^{-1}) = \begin{bmatrix} 1 & -1 \\ 0 & 1 \end{bmatrix} \begin{bmatrix} e & 0 \\ 0 & \frac{1}{e} \end{bmatrix} \begin{bmatrix} 1 & 1 \\ 0 & 1 \end{bmatrix}$$

Multiplying this all out gives us:

$$\exp(A) = \begin{bmatrix} e & \dfrac{e^2 - 1}{e} \\ 0 & \dfrac{1}{e} \end{bmatrix}$$

While this was a long example, it should show you the power of being able to calculate the functions of operators.

Summary

Well, that wraps up this chapter. I hope you can see all the wonderful decompositions you can do with matrices and some of the more advanced things you can do with them. This chapter also concludes the book. I hope you have enjoyed it and learned as much as I did. Take this math and go forth to infinity and beyond in the universe of quantum computing!

Works cited

[1] - *Triangle inequality* - Wikipedia:

`https://en.wikipedia.org/wiki/Triangle_inequality#/media/File:TriangleInequality.svg`

Section 4: Appendices

The following chapters are included in this section:

Appendix 1 Bra–ket Notation

Bra-ket notation was introduced by Paul Dirac in 1939 and is sometimes called Dirac notation consequently. Kets are denoted by a pipe ("|") and right-angle bracket ("⟩"), like so – $|\text{Label}\rangle$, while bras are denoted by a left-angle bracket ("⟨") and pipe ("|"), like so $\langle\text{Label}|$. Kets represent vectors in a Hilbert space and bras represent their covectors in a dual Hilbert space. The labels for kets and bras can be lowercase letters, numbers, and greek letters. Uppercase letters are usually reserved for operators, which we will get to later.

The computational basis vectors are represented by $|0\rangle$ and $|1\rangle$. It is important to note that the zero vector is denoted by 0 and is totally different than $|0\rangle$. The zero vector, sometimes referred to as the null vector, is the only one that is *not* represented as a ket.

An inner product between two kets, $|\phi\rangle$ and $|\psi\rangle$, is notated this way – $\langle\phi|\psi\rangle$. This can be called a "bracket" and brings the notation full circle.

Other notations used are shown in the following table:

Notation	Represents
$\|0\rangle\langle 1\|$	The outer product of kets $\|0\rangle$ and $\|1\rangle$.
$\|a\rangle \otimes \|b\rangle$	All three notations represents the tensor product of kets $\|a\rangle$ and $\|b\rangle$.
$\|a\rangle\ \|b\rangle$	
$\|ab\rangle$	
$\langle a\|B\|c\rangle$	The matrix entry. This is equal to $\langle a\| \cdot (B\|c\rangle)$.

Operators

Operators are represented by capital letters such as A, B, and C. Operators can be represented by matrices numerically, as shown in the following diagram:

$$A = \begin{bmatrix} 1 & 3 \\ 7 & 9 \end{bmatrix}$$

Notation	Represents
A^*	A complex conjugate of A
A^T	A transpose of A
$A^\dagger$	A Hermitian conjugate or adjoint of A, $A^\dagger = (A^T)^*$

The rest of bra-ket notation will be explained as the book progresses. The next section is a very advanced treatise on bras and is optional.

Bras

A bra is a linear functional. We talk about these in *Chapter 5, Transforming Space with Matrices*. To help jog your memory, they are a special case of linear transformation that takes in a vector and spits out a scalar:

$$f: V \rightarrow F \text{ where } V \text{ is a vector space and } F \text{ is the field of scalars } \mathbb{R} \text{ or } \mathbb{C}$$

For instance, I could define a linear functional for every vector in $\mathbb{R}^2$:

$$f\left(\,|v\rangle\,\right)=a+b \text{ where } |v\rangle=\begin{bmatrix} a \\ b \end{bmatrix}$$

So that:

$$f\left(\begin{bmatrix} 3 \\ 2 \end{bmatrix}\right)=3+2=5$$

$$f\left(\begin{bmatrix} 5 \\ -2 \end{bmatrix}\right)=5-2=3$$

There are many linear functionals that can be defined for a vector space. Here's another one:

$$g:\mathbb{R}^2 \to \mathbb{R}$$

$$g\left(\,|v\rangle\,\right)=2a-3b \text{ where } |v\rangle=\begin{bmatrix} a \\ b \end{bmatrix}$$

The set of all linear functionals that can be defined on a vector space actually form their own vector space called the dual vector space.

Instead of using the usual function notation for these linear functionals, Paul Dirac came up with a notation that he called a bra:

$$f_{|v\rangle} \equiv \langle v\,|$$

Since every vector has its own linear functional (called its dual vector or covector), the label between the angle bracket and vertical bar or pipe is the dual vector for the ket with the same label. In other words, every ket $|v\rangle$ has a linear functional $\langle v|$ defined for it.

Now the big question is, what is the function that is defined for each ket? That, my friend, is the inner product, defined as follows and explained in *Chapter 8, Our Place in the Universe*:

$$\left(|x\rangle, |y\rangle \right) \equiv \left(\begin{bmatrix} x_1 \\ \vdots \\ x_n \end{bmatrix}, \begin{bmatrix} y_1 \\ \vdots \\ y_n \end{bmatrix} \right) = \sum_{i=1}^{n} x_i^* y_i = x_1^* y_1 + \cdots + x_n^* y_n$$

Appendix 2: Sigma Notation

Many quantum computing books will introduce sigma notation without ever really explaining it, and you will also see shorthand for it as well. I hope to demystify this notation when you encounter it in this book and others.

Sigma

Sigma is the 18th letter of the Greek alphabet, and we are talking about the capital version, $\boldsymbol{\Sigma}$. It signifies a summation, so that:

$$x_1 + x_2 + x_3 = \sum_{i=1}^{3} x_i$$

More generally,

$$x_1 + x_2 + \cdots + x_n = \sum_{i=1}^{n} x_i$$

Here are the different parts of sigma notation:

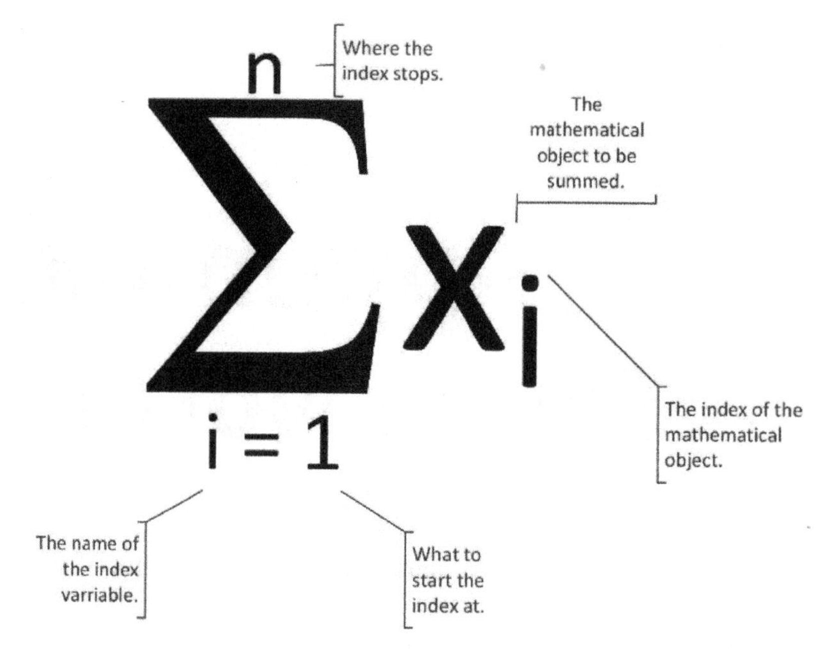

Figure 11.1 – A diagram of sigma notation

Here is an easy example to make the point better:

$$\sum_{i=1}^{3} i = 1 + 2 + 3 = 6$$

Variations

You will see shorthand in a lot of quantum computing books, but don't let it throw you. The following are all equivalent:

$$\sum_{i=1}^{n} x_i = \sum_{i} x_i = \sum x_i$$

Summation rules

There are many properties of summations, but the two most important ones I think are:

$$\sum cx_i = c\sum x_i$$

where c is a constant and

$$\left(\sum_{i=0}^{n} a_i\right)\left(\sum_{j=0}^{n} b_j\right) = \sum_{i=0}^{n}\sum_{j=0}^{n} a_i b_j$$

These are both based on the distributive property.

Appendix 3 Trigonometry

In some of the chapters, we have relied heavily on trigonometry. You probably took trig (the shortened alias for trigonometry) in high school but may have not used it for a long time. This chapter is meant as a refresher or just an introduction.

In this chapter, we are going to cover the following main topics:

- Measuring angles
- Trigonometric functions
- Formulas

Measuring angles

It all starts with angles in trigonometry. There are two main ways to measure angles – degrees and radians. Radians are almost exclusively used in quantum computing, but you are probably more familiar with degrees, so we'll start there.

Degrees

There are 360 degrees in a circle. Probably the most familiar use of degrees is in a compass, as shown in the following:

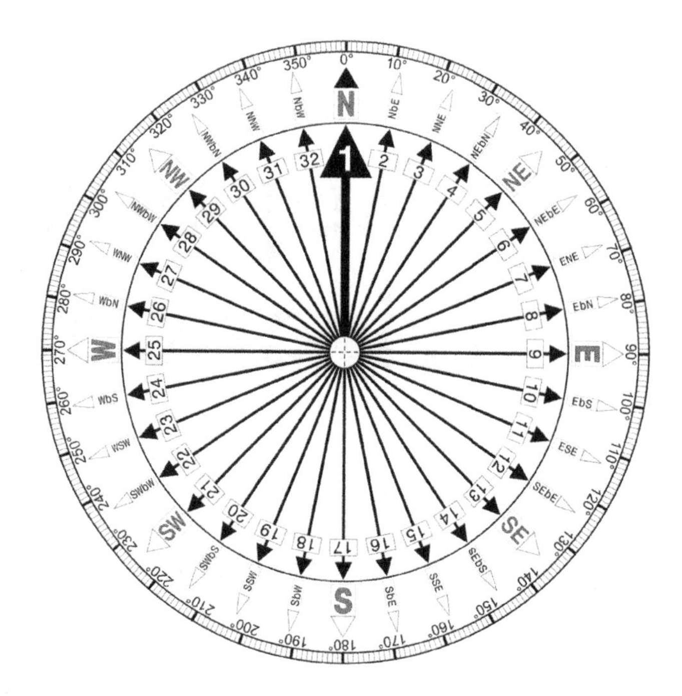

Figure 12.1 – Compass

However, on the X-Y plane, zero degrees starts on the right side of the *x* axis and goes from there. In fact, the X-Y plane can be broken up into four quadrants with each taking up 90 degrees, as shown in the following:

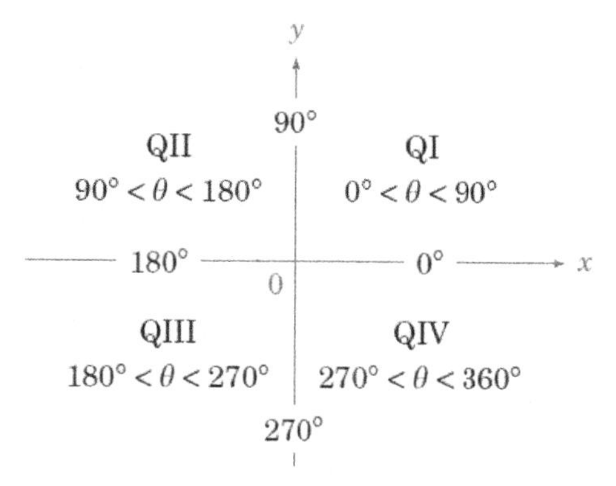

Figure 12.2 – Four quadrants [2]

Additionally, angle measures can be expressed in negative form. When degrees or radians are positive, you go in the counterclockwise direction, and when they are negative, you go in a clockwise direction on the X-Y plane, as shown in the following diagram:

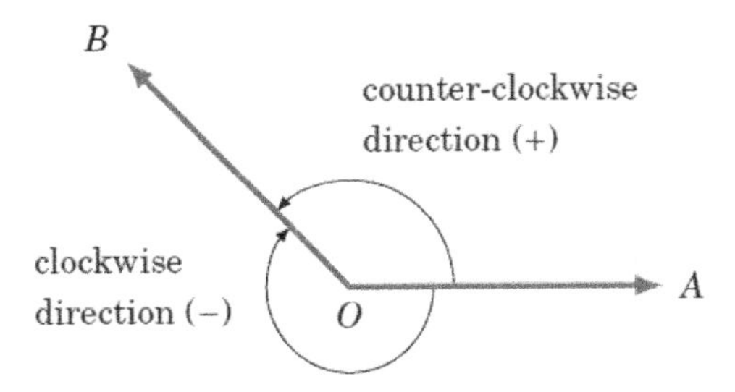

Figure 12.3 – Positive and negative angles [3]

This can lead to interesting angle measures that can be equivalently expressed in positive and negative forms. For instance, the following angle can be expressed as 45 degrees or -315 degrees:

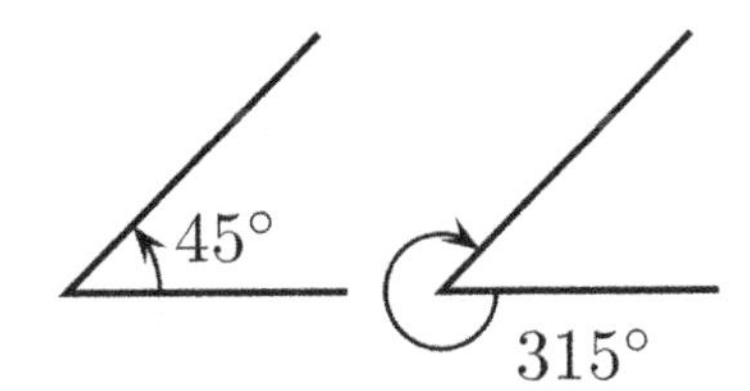

Figure 12.4 – Positive and negative angles [4]

After introducing degrees, we now get to the meat of the matter – radians!

Radians

As I said, radians are the most used measure in mathematics and physics because they make calculations much easier. The key part to converting from degrees to radians is that 360 degrees equals 2π radians. Using this, we can obtain the following formulas for converting degrees to radians:

$$1 \text{ rad} = \left(\frac{180}{\pi}\right)^{\circ} \approx 57.3^{\circ} \qquad 1^{\circ} = \left(\frac{\pi}{180} \text{ rad}\right) \approx 0.017 \text{ rad}$$

You should note that we use the abbreviation "rad" to represent a radian when we need to distinguish it from a degree.

The following diagram does a good job of showing the relationship of degrees and radians on the X-Y plane:

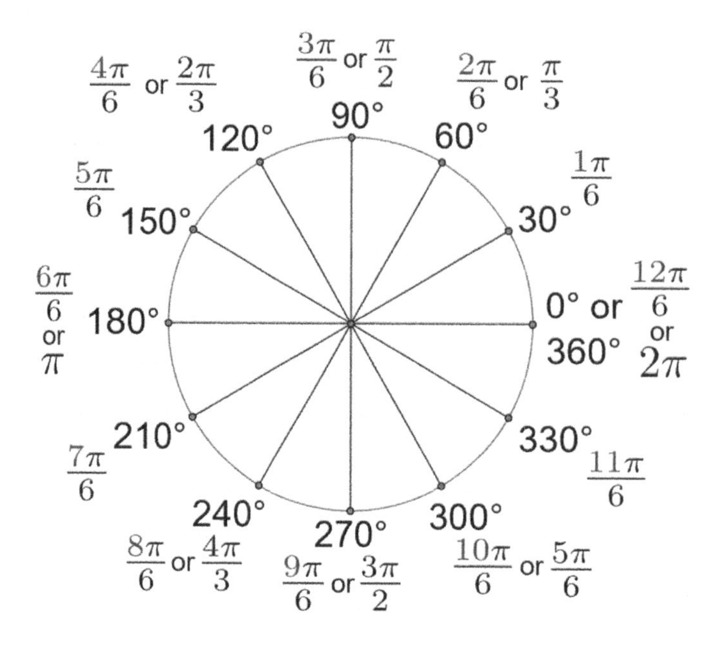

Figure 12.5 – Degrees and equivalent radians on the X-Y plane [5]

Trigonometric functions

There are three main trigonometric functions, and they are all based on right triangles. Let's use the right triangle, as follows:

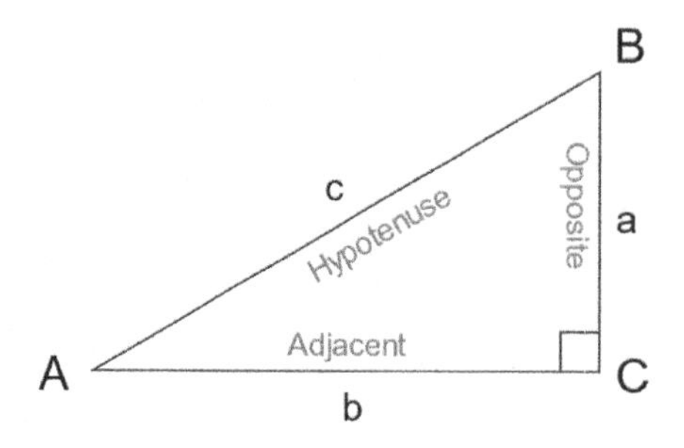

Figure 12.6 – A right triangle [6]

We will call the angle at **A** on the triangle θ. The angle at **C** is 90 degrees or a right angle. The three trig functions are called sine, cosine, and tangent. Here are their definitions along with their abbreviations:

$$\sin\theta = \frac{\text{opposite}}{\text{hypotenuse}} \qquad \cos\theta = \frac{\text{adjacent}}{\text{hypotenuse}} \qquad \tan\theta = \frac{\text{opposite}}{\text{adjacent}}$$

There are three other trig functions, which are the inverse of the main three and have the special names of cosecant, secant, and cotangent. Here are their definitions:

$$\csc\theta = \frac{\textit{hypotenuse}}{\text{opposite}} \qquad \sec\theta = \frac{\text{hypotenuse}}{\text{adjacent}} \qquad \cot\theta = \frac{\text{adjacent}}{\text{opposite}}$$

Given this, common values for these functions can be seen in the following table:

	Angle in radians (in degrees)				
Trig Function	0 (0°)	π/6 (30°)	π/4 (45°)	π/3 (60°)	π/2 (90°)
sin(θ)	0	$\frac{1}{2}$	$\frac{1}{\sqrt{2}}$	$\frac{\sqrt{3}}{2}$	1
cos(θ)	1	$\frac{\sqrt{3}}{2}$	$\frac{1}{\sqrt{2}}$	$\frac{1}{2}$	0
tan(θ)	0	$\frac{1}{\sqrt{3}}$	1	$\sqrt{3}$	undefined

Figure 12.7 – The common values for sin, cos, and tan

As angles get bigger, they move from the first quadrant into the three other quadrants and become positive or negative, based on the values of x and y in those quadrants. Here is a cheat sheet to know which are positive and negative:

QII	QI
sin +	sin +
cos −	cos +
tan −	tan +
csc +	csc +
sec −	sec +
cot −	cot +
QIII	**QIV**
sin −	sin −
cos −	cos +
tan +	tan −
csc −	csc −
sec −	sec +
cot +	cot −

Figure 12.8 – The trig functions by quadrant [7]

Formulas

There are some common formulas you should know. The first is about tangent and its relationship to the sine and cosine functions. We can derive the formula in the following way:

$$\frac{\sin\theta = \frac{\text{opposite}}{\text{hypotenuse}}}{\cos\theta = \frac{\text{adjacent}}{\text{hypotenuse}}} = \frac{\text{opposite}}{\text{hypotenuse}} \bullet \frac{\text{hypotenuse}}{\text{adjacent}} = \frac{\text{opposite}}{\text{adjacent}} = \tan\theta$$

This leads us to this:

$$\tan\theta = \frac{\sin\theta}{\cos\theta}$$

The next one we get from studying the graphs of sine and cosine as function θ over time. Here are those graphs:

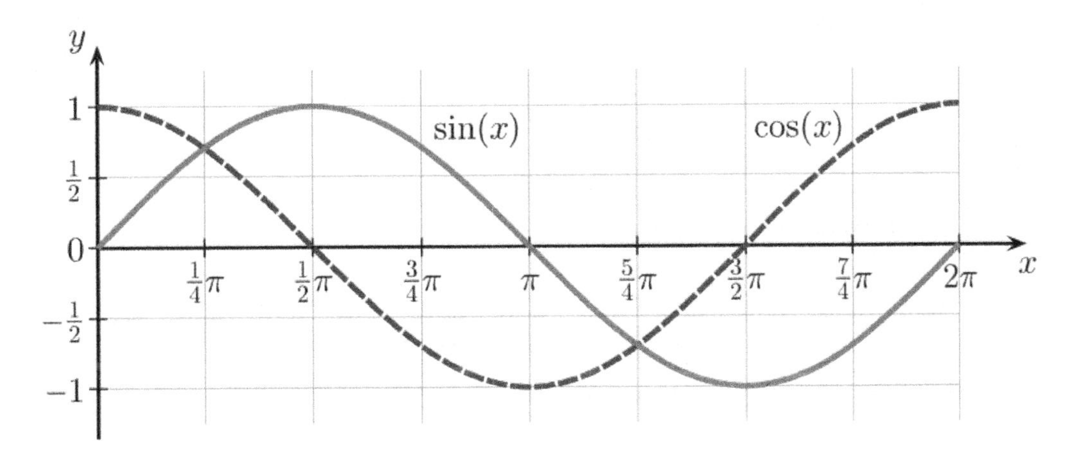

Figure 12.9 – The graphs of sine and cosine [8]

You may notice that they are "out of phase" by π/2. This leads us to the next formula:

$$\sin\theta = \cos\left(\frac{\pi}{2} - \theta\right)$$

$$\cos\theta = \sin\left(\frac{\pi}{2} - \theta\right)$$

While I will not derive it, the following is called the Pythagorean identity, and you should commit it to memory:

$$\sin^2\theta + \cos^2\theta = 1$$

Summary

Alright, that's all you need to know for trigonometry. Please consult a textbook or other resources if you feel like you need more help. The next page contains a cheat sheet of formulas you will most likely need in your journey through quantum computing.

The trig cheat sheet

Pythagorean identities

$$\cos^2(x) + \sin^2(x) = 1 \quad \sec^2(x) - \tan^2(x) = 1$$
$$\csc^2(x) - \cot^2(x) = 1$$

Double angle identities

$$\sin(2x) = 2\sin(x)\cos(x) \qquad \cos(2x) = 1 - 2\sin^2(x)$$
$$\cos(2x) = 2\cos^2(x) - 1 \quad \cos(2x) = \cos^2(x) - \sin^2(x)$$
$$\tan(2x) = \frac{2\tan(x)}{1 - \tan^2(x)}$$

Sum/difference identities

$$\sin(s+t) = \sin(s)\cos(t) + \cos(s)\sin(t)$$
$$\sin(s-t) = \sin(s)\cos(t) - \cos(s)\sin(t)$$
$$\cos(s+t) = \cos(s)\cos(t) - \sin(s)\sin(t)$$
$$\cos(s-t) = \cos(s)\cos(t) + \sin(s)\sin(t)$$
$$\tan(s+t) = \frac{\tan(s) + \tan(t)}{1 - \tan(s)\tan(t)}$$
$$\tan(s-t) = \frac{\tan(s) - \tan(t)}{1 + \tan(s)\tan(t)}$$

Product-to-sum identities

$$\cos(s)\cos(t) = \frac{\cos(s-t)+\cos(s+t)}{2}$$

$$\sin(s)\sin(t) = \frac{\cos(s-t)-\cos(s+t)}{2}$$

$$\sin(s)\cos(t) = \frac{\sin(s+t)+\sin(s-t)}{2}$$

$$\cos(s)\sin(t) = \frac{\sin(s+t)-\sin(s-t)}{2}$$

Works cited

[1] - File:Compass Card B+W.svg – Wikimedia Commons (https://commons.wikimedia.org/wiki/File:Compass_Card_B%2BW.svg)

[2] - 8.png (438×315) (opencurriculum.org) (http://media.opencurriculum.org/articles_manual/michael_corral_trigonometry/trigonometric-functions-of-any-angle/8.png)

[3] - Trigonometric Functions of Any Angle – OpenCurriculum (2.png) (https://opencurriculum.org/5484/trigonometric-functions-of-any-angle/)

[4] - File:Positive, negative angle.svg – Wikimedia Commons (https://commons.wikimedia.org/wiki/File:Positive,_negative_angle.svg)

[5] - File:30 degree rotations expressed in radian measure.svg – Wikimedia Commons (https://commons.wikimedia.org/wiki/File:30_degree_rotations_expressed_in_radian_measure.svg)

[6] - File:TrigonometryTriangle.svg – Wikimedia Commons (https://commons.wikimedia.org/wiki/File:TrigonometryTriangle.svg)

[7] - Trigonometric Functions of Any Angle – OpenCurriculum (9.png) (https://opencurriculum.org/5484/trigonometric-functions-of-any-angle/)

[8] - File:Sine cosine one period.svg – Wikimedia Commons (https://commons.wikimedia.org/wiki/File:Sine_cosine_one_period.svg)

Appendix 4 Probability

The study of gambling, specifically the throwing of dice, led to the mathematical field of probability. Probability is the study of how likely an event is to occur given a number of possible outcomes. A real number between 0 and 1 is assigned to each event where 0 signifies the event has no chance of happening and 1 signifies the event will always happen. You can also multiply these numbers by 100 to get a percentage that the event will happen. All the probabilities for all possible outcomes must sum to 1. For instance, the probability of a coin flip landing on heads is 0.5 or 50%. For tails, it is also 0.5 or 50%. Both of these numbers add up to 1. From these basics, this chapter will go over the probability needed in the study of quantum computing.

In this chapter, we are going to cover the following main topics:

- Definitions
- Random variables

Definitions

Let's start by getting some basic definitions out of the way. The word **experiment** is used in probability theory to denote the execution of a procedure that produces a random outcome. Examples of experiments are flipping a coin or rolling dice. In quantum computing, an experiment is measuring a qubit.

A **sample space** is the set of all possible outcomes of an experiment. It is usually denoted by Ω (the upper case Greek letter omega). The set Ω for a fair coin is {Heads, Tails}. The set Ω for one die is {1, 2, 3, 4, 5, 6}. The set Ω for a qubit when measured in the Z basis is $\{|0\rangle, |1\rangle\}$.

An **event** (E) is a subset of Ω. Every outcome is a subset of size 1 – for example, {Heads} and {Tails} are events for a fair coin. But as we saw in *Chapter 3, Foundations*, subsets also include the empty set Ø and the whole set itself, which is Ω in this case. The set of all events is called an event space and is usually denoted with a $\mathcal{F}$. Here is the event space for a fair coin:

- Ø
- {Heads}
- {Tails}
- {Heads, Tails} = Ω

Because of the definition of an event, you can also group outcomes together. For instance, {1 ,2, 3} is the event that a die is 3 or below after a roll (aka experiment).

Finally, there is a probability function (P) that maps each event to a real number between 0 and 1:

$$P : F \rightarrow [0,1]$$

The probability function has these properties:

$$P(\Omega) = 1$$
$$P(\varnothing) = 0$$
$$\text{For all E such that } E \subseteq \Omega$$
$$0 \le P(E) \le 1$$

Here is an example for a flip of a coin using these definitions, where H stands for Heads and T stands for Tails:

$$\Omega = \{H, T\}$$
$$F = \{\varnothing, \{H\}, \{T\}, \{H, T\}\}$$
$$P(\varnothing) = 0$$
$$P(\{H\}) = \frac{1}{2}$$
$$P(\{T\}) = \frac{1}{2}$$
$$P(\{H, T\}) = 1$$

Let's move on to to see how we can analyze these events further.

Random variables

An important concept in probability is a **random variable**. Oftentimes, we are not interested in the actual outcome of an experiment but some function of the outcome. For instance, let's define the function S to be the number of tails when flipping two coins. We know that Ω is {(H,H), (H,T), (T,H), (T,T)} where H stands for heads and T stands for tails. We also know that the probability of each of these outcomes is $1/4^{th}$. However, I want to know the amount of tails in my outcomes, which I define as this:

$$S : \Omega \rightarrow \mathbb{R}$$
$$S(HH) = 0$$
$$S(HT) = 1$$
$$S(TH) = 1$$
$$S(TT) = 2$$

If I define S to be a random variable, then:

$$P(S = 0) = \frac{1}{4}$$
$$P(S = 1) = \frac{1}{2}$$
$$P(S = 2) = \frac{1}{4}$$

In general, random variables are written with capital letters such as *X*, *Y*, and *Z*. Random variables are functions from the sample space Ω to a measurable space, which is not a trivial thing. Fortunately for us, for most of our random variables, the measurable space will be the real numbers.

Discrete random variables

There are continuous and discrete random variables. Most random variables in quantum computing are discrete, and hence, we will only deal with this type. Discrete random variables have distinct values, and their sample spaces are finite or countably infinite. Instead of writing $P(X = z)$ all the time, we define a new function called the **Probability Mass Function** (**PMF**) this way:

$$p(z) = P(X = z)$$

Two properties of the PMF are:

$$\sum_{i=1}^{n} p(x_i) = 1, \text{ where n is the amount of possible values}$$

$$p(x) \geq 0$$

Histograms are often used to show the PMF graphically for a random variable. Here is a nice histogram showing the PMF for a random variable *S*, defined as the sum of two dice being rolled:

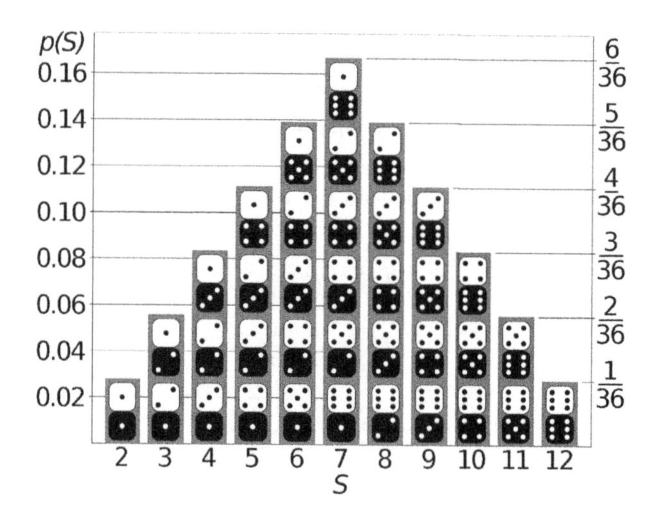

Figure 13.1 – The PMF of S, the sum of two dice rolled

Let's look at how we can describe our random variables.

The measures of a random variable

There are a few measures that are important to know for random variables. The first is the **expected value** of a random variable denoted as *E[X]* where *X* is the random variable. Let's start with an example first and look at the roll of a single die. The expected value takes each possible outcome and multiplies it by the PMF for that value. So, if we let *X* represent the value of the outcome of a die roll, then the expected value will be:

$$p(1) = p(2) = p(3) = p(4) = p(5) = p(6) = \frac{1}{6}$$

$$E[X] = 1 \cdot p(1) + 2 \cdot p(2) + 3 \cdot p(3) + 4 \cdot p(4) + 5 \cdot p(5) + 6 \cdot p(6)$$

$$E[X] = 1 \cdot \frac{1}{6} + 2 \cdot \frac{1}{6} + 3 \cdot \frac{1}{6} + 4 \cdot \frac{1}{6} + 5 \cdot \frac{1}{6} + 6 \cdot \frac{1}{6}$$

$$E[X] = \frac{1}{6} + \frac{2}{6} + \frac{3}{6} + \frac{4}{6} + \frac{5}{6} + \frac{6}{6} = \frac{21}{6} = \frac{7}{2} = 3.5$$

$$E[X] = 3.5$$

You should note that the expected value is not actually a possible value of the PMF. This will often be true. Intuitively, this should also look like the **average** or **mean**, and it is if all outcomes have the same PMF value. Let's define this mathematically. The expected value for a random variable *X* is defined as:

$$E[X] \equiv \sum_{i=1}^{n} x_i p(x_i)$$

The other important measure of a random variable is its **variance**, which measures the spread of possible PMF values. It is defined as:

$$\text{var}(X) \equiv E[(X - E[X])^2] = E[X^2] - E(X)^2$$

Let's calculate the variance for our roll of one die example. We already know the expected value of X, but we also need the expected value of X^2. Here's the calculation:

$$p(1) = p(2) = p(3) = p(4) = p(5) = p(6) = \frac{1}{6}$$

$$E[X^2] = 1^2 \cdot p(1) + 2^2 \cdot p(2) + 3^2 \cdot p(3) + 4^2 \cdot p(4) + 5^2 \cdot p(5) + 6^2 \cdot p(6)$$

$$E[X^2] = 1 \cdot \frac{1}{6} + 4 \cdot \frac{1}{6} + 9 \cdot \frac{1}{6} + 16 \cdot \frac{1}{6} + 25 \cdot \frac{1}{6} + 36 \cdot \frac{1}{6}$$

$$E[X^2] = \frac{1}{6} + \frac{4}{6} + \frac{9}{6} + \frac{16}{6} + \frac{25}{6} + \frac{36}{6} = \frac{91}{6}$$

$$E[X^2] = \frac{91}{6}$$

Now, we can calculate the variance of our roll of one die:

$$\text{var}(X) = E[X^2] - E(X)^2$$

$$\text{var}(X) = \frac{91}{6} - \left(\frac{7}{2}\right)^2 = \frac{35}{12}$$

The last measure to consider is called the **standard deviation** of a random variable, and it is quite easy once you have the variance. It is just the square root of the variance.

And there you go – you now know the three most important measures of a random variable: the expected value, the variance, and the standard deviation.

Summary

The field of probability is vast, but you now have the necessary tools to understand it as it applies to quantum computing. The foundation for understanding probability is the definitions we went through, and in quantum computing, it all revolves around random variables.

Works cited

[1] - File:Dice Distribution (bar).svg - Wikimedia Commons (`https://commons.wikimedia.org/wiki/File:Dice_Distribution_(bar).svg`)

Appendix 5 References

"Abstract Algebra." *YouTube*, uploaded by Socratica, January 2, 2021, www.youtube.com/c/Socratica.

Andreescu, Titu, and Andrica, Dorin. *Complex Numbers from A to...Z. Birkhäuser Boston*, 2014.

Axler, Sheldon. *Linear Algebra Done Right (Undergraduate Texts in Mathematics).* 3rd edition. 2015, *Springer*, 2014.

Byron, Frederick, and Fuller, Robert. *Mathematics of Classical and Quantum Physics. Dover Publications*, 1992.

Dirac, P. A. M. "A New Notation for Quantum Mechanics." *Mathematical Proceedings of the Cambridge Philosophical Society*, vol. 35, no. 3, 1939, pp. 416–18. *Crossref*, https://doi.org/10.1017/s0305004100021162.

Encyclopedia of Mathematics. Encyclopedia of Mathematics community, encyclopediaofmath.org/wiki/Main_Page. Accessed 1 Feb. 2021.

Gowers, Timothy, et al. *The Princeton Companion to Mathematics. Amsterdam University Press*, 2008.

Greenfield, Pavel. "Linear Algebra." *YouTube*, uploaded by *MathTheBeautiful*, 4 Jan. 2021, www.youtube.com/c/MathTheBeautiful.

Griffiths, David. *Introduction to Quantum Mechanics.* 2nd edition, *Cambridge University Press*, 2016.

Halmos, Paul. *Naive Set Theory (Dover Books on Mathematics).* Reprint, *Dover Publications*, 2017.

Hidary, Jack. *Quantum Computing: An Applied Approach.* 1st edition 2019, *Springer*, 2019.

Lay, David, et al. *Linear Algebra and Its Applications.* 5th edition, *Pearson*, 2014.

"Math on YouTube." *YouTube*, uploaded by *Eddie Woo*, 2 Jan. 2021, `www.youtube.com/c/misterwootube`.

McMahon, David. *Quantum Computing Explained. Wiley*, 2007.

Nielsen, Michael A., and Chuang, Isaac. *Quantum Computation and Quantum Information: 10th Anniversary Edition.* 1st edition, *Cambridge University Press*, 2011.

Pinter, Charles. *A Book of Abstract Algebra: Second Edition (Dover Books on Mathematics).* Second, *Dover Publications*, 2010.

"Professor M does Science." *YouTube*, uploaded by Professor M, 2 Jan. 2021, `www.youtube.com/c/ProfessorMdoesScience`.

Rieffel, Eleanor, and Polak, Wolfgang. *Quantum Computing: A Gentle Introduction (Scientific and Engineering Computation).* Illustrated, *The MIT Press*, 2014.

Ross, Sheldon. *A First Course in Probability. Prentice Hall*, 1998.

Sanderson, Grant. Various math videos. *YouTube*, uploaded by 3Blue1Brown, 1 Jan. 2020, `www.youtube.com/c/3blue1brown`.

Shankar. *Principles of Quantum Mechanics.* 2nd edition, *Plenum Press*, 1994.

Stewart, Ian, and Tall, David. *The Foundations of Mathematics.* 2nd edition, *Oxford University Press*, 2015.

Stewart, James. *Single Variable Calculus: Early Transcendentals, Volume I.* 8th edition, *Cengage Learning*, 2015.

Strang, Gilbert. *Introduction to Linear Algebra (Gilbert Strang).* 5th edition, *Wellesley-Cambridge Press*, 2016.

Susskind, Leonard, and Friedman, Art. *Quantum Mechanics: The Theoretical Minimum.* Illustrated, *Basic Books*, 2015.

Sutor, Robert. *Dancing with Qubits: How Quantum Computing Works and How It Can Change the World. Packt Publishing*, 2019.

Weisstein, Eric. "Wolfram MathWorld: The Web's Most Extensive Mathematics Resource." *Wolfram MathWorld*, Wolfram Research, `mathworld.wolfram.com`. Accessed 1 Feb. 2021.

Wikipedia contributors. Various articles. *Wikipedia*, `en.wikipedia.org/wiki/Main_Page`. Accessed 1 Feb. 2021.

Index

A

B

C

D

E

F

G

Q

R

S

T

U

V

W

Other Books You May Enjoy

If you enjoyed this book, you may be interested in these other books by Packt:

Quantum Computing with Silq Programming

Srinjoy Ganguly, Thomas Cambier

ISBN: 978-1-80056-966-9

- Identify the challenges that researchers face in quantum programming
- Understand quantum computing concepts and learn how to make quantum circuits
- Explore Silq programming constructs and use them to create quantum programs
- Use Silq to code quantum algorithms such as Grover's and Simon's
- Discover the practicalities of quantum error correction with Silq
- Explore useful applications such as quantum machine learning in a practical way

Learn Quantum Computing with Python and IBM Quantum Experience

Robert Loredo

ISBN: 978-1-83898-100-6

- Explore quantum computational principles such as superposition and quantum entanglement
- Become familiar with the contents and layout of the IBM Quantum Experience
- Understand quantum gates and how they operate on qubits
- Discover the quantum information science kit and its elements such as Terra and Aer
- Get to grips with quantum algorithms such as Bell State, Deutsch-Jozsa, Grover's algorithm, and Shor's algorithm
- How to create and visualize a quantum circuit

Packt is searching for authors like you

If you're interested in becoming an author for Packt, please visit `authors.packtpub.com` and apply today. We have worked with thousands of developers and tech professionals, just like you, to help them share their insight with the global tech community. You can make a general application, apply for a specific hot topic that we are recruiting an author for, or submit your own idea.

Share Your Thoughts

Now you've finished *Essential Mathematics for Quantum Computing*, we'd love to hear your thoughts! Scan the QR code below to go straight to the Amazon review page for this book and share your feedback or leave a review on the site that you purchased it from.

`https://packt.link/r/1-801-07314-7`

Your review is important to us and the tech community and will help us make sure we're delivering excellent quality content.

Made in the USA
Coppell, TX
01 December 2022